HYDRODYNAMICS AND MASS TRANSFER IN DOWNFLOW SLURRY BUBBLE COLUMNS

HYDRODYNAMICS AND MASS TRANSFER IN DOWNFLOW SLURRY BUBBLE COLUMNS

Subrata Kumar Majumder, PhD

Apple Academic Press Inc.
3333 Mistwell Crescent
Oakville, ON L6L 0A2
Canada

Apple Academic Press Inc.
9 Spinnaker Way
Waretown, NJ 08758
USA

Library and Archives Canada Cataloguing in Publication

Majumder, Subrata Kumar, author
Hydrodynamics and mass transfer in downflow slurry bubble columns /
Subrata Kumar Majumder, PhD.

Includes bibliographical references and index.
Issued in print and electronic formats.
ISBN 978-1-77188-673-4 (hardcover).--ISBN 978-1-351-24986-7 (PDF)

1. Gases--Absorption and adsorption. 2. Chemical reactions.
3. Hydrodynamics. 4. Mass transfer. I. Title.

| TP156.A3M35 2018 | 660'.28423 | C2018-902277-9 | C2018-902278-7 |

CIP data on file with US Library of Congress

ABOUT THE AUTHOR

Subrata Kumar Majumder, PhD

Subrata Kumar Majumder, PhD, is a professor in the Chemical Engineering Department at the Indian Institute of Technology Guwahati, India. His research interests include multiphase flow and reactor development, hydrodynamics in multiphase flow, mineral processing, and process intensification and micro–nano bubble science and technology and its applications. He is a recipient of several awards and serves on the editorial boards of several journals. He has authored five book chapters, has presented 27 conference papers and has published 74 articles in several reputed refereed journals. He has carried out several sponsored and consultancy projects and has collaborated internationally. Presently, he is working in the field of microbubble science and technology and its applications in mineral beneficiation, removal of arsenic, ammonia, dye, phenolic and pharmaceutical derivatives and process intensification by developing ejector-induced gas-aided extraction process.

CONTENTS

LIST OF ABBREVIATIONS

2D	two dimensional
ABND	average bubble number density
AC	alternating current
ADM	axial dispersion model
CFD	computational fluid dynamics
CFL	compact fluorescent lamp
DGISR	downflow gas-interacting slurry reactor
DQMOM	direct quadrature method of moments
EMD	empirical mode decompositions
FCC	fluid catalytic cracking
LPG	liquefied petroleum gas
MUSIG	multiple size group
PBA	particle-to-bubble adhesion
PBE	population balance equation
QMOM	quadrature method of moments
RTD	residence time distribution
TCP	tricalcium phosphate

LIST OF SYMBOLS

$\dot{E}_v$ Average energy dissipation rate per unit volume (N·m/m³)

u_* Friction velocity (m/s)

$\hat{t}$ Time ratio ($= t_i / t_m$) (-)

$\in$ Average energy dissipation rate per unit volume (N·m/m³)

α Angle of jet inclination (degree)

β Coefficient related to the boundary layer thickness (-)

ξ Curvature parameter ($= 8[v_g L_j / (u_o d_j^2)]^{1/2}$) (-)

λ Parameter (-)

Γ Parameter (-)

β Parameter (-)

σ Surface tension (N/m)

θ_c Critical angle (degree)

v_g Gas kinematic viscosity (m²/s)

α_g, ε_g Fractional gas volume fraction (-)

ϕ_g, ϕ_{sl} Lockhart–Martinelli parameters (-)

v_l Kinematic viscosity of liquid (μ/ρ)

ε_1 Liquid holdup (-)

δ_1 Thickness of boundary layer (m)

ΔP Pressure drop (N/m²)

ΔP_a Acceleration component of pressure drop (Pa)

ΔP_{fog} Gas phase frictional pressure drop (Pa)

ΔP_{fosl} Slurry phase frictional pressure drop (Pa)

ΔP_{ftp} Three-phase frictional pressure drop (Pa)

ΔP_h Hydrostatic head (Pa)

ΔP_{tp} Total pressure drop for three-phase flow (Pa)

Δz Length of the column across which pressure drop is measured (m)

μ_{sl} Slurry viscosity (Pa·s)

A Interfacial area (m²)

a Parameter (-)

a Specific interfacial area (1/m)

A_c Cross-sectional area of the column (m²)

A_d Displacement area (m²)

A_e Area of the electrode reacting surface (m²)

A_j Cross-sectional area of the jet (m^2)

Ar Archimedes number ($d_p^3 \rho_{sl} (\rho_p - \rho_{sl}) \, g / \mu_{sl}^2$) (-)

b Area ratio of jet to column (-)

C Concentration (mol/m^3)

C Parameter used in the Equations 4.16 and 4.17 (-)

C^* Equilibrium concentration (mol/m^3)

C_1^* Equilibrium concentration in the liquid phase (mol/m^3)

$C_{1,hi}^*$ Equilibrium concentration at section of height h_i (mol/m^3)

Ca Capillary number (vu/σ) (-)

C_B Fraction of the bubble-generated turbulent energy (-)

C_{B0} Concentration of B in bulk liquid phase (mol/m^3)

C_f Frictional drag coefficient (-)

$C_i(t)$ Tracer concentration at time t (mol/m^3)

C_i Concentration of reacting ion solution (mol/m^3)

C_l Molar concentration of solute in liquid (mol/m^3)

$C_{l,hi}$ Concentration of liquid at section of height h_i (mol/m^3)

C_m Energy dissipation coefficient of mixing (-)

c_o Input tracer concentration (mol/m^3)

c_o Distribution parameter (-)

C_o Initial concentration (mol/m^3)

C_s Concentration of ion at electrode surface (mol/m^3)

C_T Craya–Curtet number ($\sqrt{(d_c / d_j)^2 - 1/2}$) (-)

C_t Tracer concentration at time t (mol/m^3)

C_v Slurry concentration (vol.%)

C_{vm} Virtual mass coefficient (-)

D_A^* Diffusivity of component A at equilibrium condition (m^2/s)

d_{32} Sauter mean bubble diameter (m)

d_b Bubble diameter (m)

D_b Dispersion coefficient of bubble motion (m^2/s)

d_{be} Equivalent diameter of the bubble (m)

d_{bi} Bubble diameter at ith class (m)

d_c Diameter of the vertical column (m)

d_c Diameter of the column (m)

d_c Downcomer or column diameter (m)

d_e Electrode diameter (m)

d_j Jet diameter (m)

d_j Nozzle diameter (m)

D_1 Diffusivity of reacting species (m^2/s)

d_{max}	Maximum bubble diameter (m)
d_{min}	Minimum bubble diameter (m)
d_n	Nozzle diameter (m)
d_o	Orifice hole diameter (m)
d_p	Particle diameter (m)
D_{pq}	Bubble mean diameter at p and q classes (m)
d_r	Jet diameter at the reference point (m)
D_r	Diameter ratio of column to particle (d_c/d_p) (-)
D_r	Diameter ratio of nozzle to particle (d_n/d_p) (-)
$Dr._{,32}$	Ratio of mean bubble diameter to particle diameter (d_{32}/d_p) (-)
d_s/d_t	Transfer rate of ionic species (-)
D_{sl}	Dispersion coefficient (m²/s)
D_T	Particle diffusion coefficient (m²/s)
E	Applied potential (v)
E	Exchange factor (-)
E_{bt}	Bubble-generated turbulent energy (N·m)
E_f	Kinetic energy loss due to friction (N·m)
e_m	Energy dissipation rate of mixing per unit volume (-)
E_{min}	Minimum energy (W)
E_{mt}	Rate of energy transfer from mean to turbulent flow (N·m)
E_s	Kinetic energy (W)
E_s	Supplied kinetic energy (N·m)
E_u	Kinetic energy utilized (N·m)
E_{um}	Energy utilized (W)
E_z	Dispersion coefficient (m²/s¹)
ε	Dissipation of turbulent kinetic energy (N·m)
ε_g	Gas holdup (-)
$f(x)$	Distribution function
F	Faraday constant (C)
F_f	Frictional drag force (N)
F_{lt}	Fraction of light transmission (-)
f_o	Friction factor for single-phase flow (-)
Fr	Froude number based on column diameter (V_M^2/gd_c) (-)
Fr_j	Froude number (u_j^2/gd_o) (-)
F_{rp}	Froude number based particle diameter (u_{sl}^2/gd_p) (-)
f_{tpl}	Friction factor for liquid in three-phase flow (-)
g	Acceleration due to gravity (m/s²)
H(t)	Information entropy (-)

H_c	Column height (m)
h_i	Height of the column at ith section
h_m	Gas–liquid–solid mixing height in the column (m)
h_m	Gas–slurry–solid mixing height (m)
h_m	Gas–solid–liquid mixing height (m)
h_m	Phase mixture height (m)
$h_{r,i}$	Dimensionless distance h_i / h_m (-)
h_{sl}	Clear liquid–solid height in the column (m)
h_{sl}	Clear liquid–solid slurry height (m)
i	Current (A)
i_d	Limiting current density (A/m²)
$I_i(t)$	Information amount (-)
i_l	Limiting current (A)
k	Rate constant (mol/s·(mol/m³)ⁿ)
k	Reaction rate constant
K	Velocity characteristic factor (-)
k_l	Mass transfer coefficient (m/s)
L	Length (m)
l	length of the cylindrical section of a nozzle (m)
L_B	breakup length of a liquid jet (m)
LB	Lower bound
L_j	Jet length (m)
L_m	Length of the mixing zone (m)
LPG	Liquefied petroleum gas (-)
M (t)	Quality of mixedness (-)
m	Coefficient (-)
M	Mass scale of gas phase (bubble)
M	Parameter, molecular weight (-)
m^k	Moments of particle (bubble) size distribution
Mo	Morton number, $(= g\mu_{SL}^4 / \rho_{SL}\sigma_{SL}^3)$ (-)
M_r	Mass ratio of gas to slurry (-)
$\dot{m}_{sl}$	Mass flow rate of solid–liquid slurry (kg/s)
n	coefficient (-)
n_b	Number of bubbles (-)
N_i	Bubble number frequency in class i
n_i	Number of bubbles at class i (-)
p	Parameter (-)
P	Parameter (m)

P	Pressure (Pa)
P_{atm}	Atmospheric pressure (Pa)
P_b	Pressure fluctuation generated by a single moving bubble in an infinite medium
P_c	Probability that a collision
Pe	Peclet number $(u_{sl}z / D_{sl})$ (-)
P_{hi}	Pressure at section of height h_i (N/m²)
$P_i(t)$	Probability of tracer concentration in semicylindrical cell (-)
ppi	Pores per inch (-)
P_s	Separator pressure drop (N/m²)
P_t	Total column pressure (N/m²)
q	Parameter (-)
$Q_{eddy,max}$	The maximum flow rate of the recirculating eddy (m/s)
Q_g	Volumetric flow rate of gas (m³/s)
Q_g	Volumetric flow rate of gas (kg/m³)
Q_{sl}	Slurry flow rate (m³/s)
Q_{sl}	Volumetric flow rate of slurry (kg/m³)
Q_{sl}	Volumetric flow rate of solid and liquid (m³/s)
Q_{slm}	Minimum volumetric flow rate of solid and liquid (m³/s)
r	Radius of the reactor (m)
R_A	Absorption rate per unit interfacial area (mol/m²s)
Re	Reynolds number based on slurry properties $(d_c u_{sl} \rho_{sl} / \mu_{sl})$ (-)
Re_g	Gas Reynolds number $(= d_c u_g \rho_g / \mu_g)$ (-)
Re_{length}	Reynolds number $(u_o L_j / v_l)$ (-)
Re_n	Reynolds number based on nozzle diameter $(d_n u_j \rho_{sl} / \mu_{sl})$ (-)
Re_o	Reynolds number based on single phase (-)
Re_{sl}	Reynolds number based on slurry velocity $(d_c u_{sl} \rho_{sl} / \mu_{sl})$ (-)
Re_{sl}	Reynolds number of slurry flow (-)
Re_{sl}	Slurry Reynolds number $(= d_c u_{sl} \rho_{sl} / \mu_{sl})$ (-)
Re_{slb}	Bubble Reynolds number $(= \rho_{sl} d_b u_{sl} / \mu_{sl})$ (-)
Re_{slb}	Reynolds number of slurry based on bubble diameter (-)
Re_{slp}	Reynolds number based on particle diameter (-)
r_n	Nozzle radius (m)
S_d	Cross-sectional area of downflow motive bubble (m²)
S_u	Cross-sectional area of upflow motive bubble (m²)
T	Temperature (°C)
t	Time (s)
T_c	Gas film thickness (m)

t_i^+	Transport number of reacting species (-)
t_m	Mean residence time (s)
u_0	Relative velocity of the two bubbles at the onset of deformation (m/s)
u_b	Bubble rise velocity (m/s)
u_b	Bubble terminal rise velocity (m/s)
UB	Upper bound
u_c	Liquid circulation velocity (m/s)
u_d	Drift velocity (m/s)
$u_{eddy,max}$	The maximum velocity of the recirculating eddy (m/s)
u_g	Actual gas velocity (m/s)
u_g	Local velocity of boundary layer gas (m/s)
u_g	Superficial velocity of the gas in the column (m/s)
u_{gd}	Gas velocity at diffuser (m/s)
u_j	Jet velocity (m/s)
u_{jm}	Minimum jet velocity (m/s)
u_l	Actual liquid velocity at the column centre (m/s)
u_l	Superficial liquid velocity (m/s)
u_m	Gas–liquid–solid mixture velocity (m/s)
u_m	Gas–slurry–solid mixture velocity (m/s)
u_m	Mixture velocity (m/s)
u_o	Jet velocity related to the nozzle outlet (m/s)
u_o	Velocity at the column axis (m/s)
U_r	Ratio of slurry velocity to gas velocity (u_{sl}/u_g) (-)
u_s	Slip velocity (m/s)
u_{sg}	Superficial gas velocity (m/s)
u_{sl}	Interstitial slurry velocity (m/s)
u_{sl}	Slurry superficial velocity in the column (m/s)
u_{sl}	Superficial liquid velocity (m/s)
u_{sl}	Superficial Slurry velocity (m/s)
u_{sl}	Superficial solid–liquid slurry velocity (m/s)
u_{slm}	Minimum slurry velocity (m/s)
u_t	Turbulent velocity (m/s)
V_D	Dispersion volume (m³)
V_g	Volume of gas phase (m³)
V_i	Volume of the semicylindrical cell (m³)
V_r	Velocity ratio of slurry to gas (u_{sl}/u_g) (-)
w	Slurry concentration (%)

w	Slurry concentration (gram of solids per 100 ml of liquid) (%)
w	Slurry concentration (wt.%)
We	Weber number $(u_j^2 d_n \rho_{sl}/\sigma_{sl})$ (-)
We_c	Critical Weber number (-)
We_g	Weber number $(u_o^2 d_j \rho_g / \sigma)$ (-)
We_o	Weber number $(u_o^2 d_n \rho_l / \sigma)$ (-)
X	Lockhart–Martinelli parameter (-)
x	Variable
x	Horizontal coordinate (m)
X	Lockhart–Martinelli parameter (-)
x	Quality (-)
X_{mod}	Modified Lockhart–Martinelli parameter (-)
y	Vertical coordinate (m)
Z	Ohnesorge number $(\mu_l / \sqrt{\rho_l \sigma L_j})$ (-)
z	Axial position (m)
z	Dispersion height (m)
z	Number of electrons released or consumed during the reaction (-)
α	Parameter (-)
α_{max}	Maximum allowable void fraction (-)
β	Breakup kernel constant (-)
β	Parameter (-)
γ	Ratio of hydrostatic pressure to the total column pressure (-)
δ	Thickness of diffusion layer (-)
ε_g	Gas holdup (-)
η_{bc}	Mass transfer efficiency of the column (-)
η_m	Energy efficiency (-)
η_m	Energy utilization efficiency (-)
λ	Constant (-)
λ	Gas-to-liquid volumetric flow ratio (-)
μ_{eff}	Effective viscosity (Pa·s)
μ_g	Gas viscosity (Pa·s)
μ_g	Viscosity of the gas (kg/m·s)
μ_l	Liquid viscosity (Pa·s)
μ_{sl}	Slurry viscosity (N·s/m²)
μ_{sl}	Slurry viscosity (Pa·s)
μ_{sl}	Slurry viscosity (kg/m·s)
μ_{sl}	Solid–liquid slurry viscosity (Pa·s)
ρ_g	Density of gas (kg/m³)

ρ_l	Density of liquid (kg/m^3)
ρ_m	Density of mixture (kg/m^3)
ρ_p	Density of particle (kg/m^3)
ρ_s	Density of solid (kg/m^3)
ρ_{sl}	Density of liquid–solid slurry (kg/m^3)
ρ_{sl}	Density of slurry (kg/m^3)
σ	Surface tension (N/m)
σ^2	Variance (-)
σ_l	Liquid surface tension (N/m)
σ_{sl}	Slurry surface tension (N/m)
τ_p	Response time of measuring electrode (s)
χ	Henry's law constant (-)

ABOUT THE BOOK

This reference book is intended for process intensification of three-phase flow, not only in chemical engineering but also in related disciplines of biochemical engineering and mechanical engineering. The interest in applications of a gas interacting downflow slurry bubble column reactor (GIDSBCR) for process intensifications of chemical and biochemical industries is due to the following advantages: the bubbles are finer and more uniform in size, coalescence of bubbles are negligible, homogenization of the two phases in the whole column is possible, a large amount of liquid can be contacted with a small amount of dispersed gas efficiently, and there is a higher residence time of the gas bubbles, a lower power consumption, almost complete gas utilization, higher overall mass transfer coefficient, and tolerance to particulates. It is, therefore, useful for slurry chemical reactions. Slurry reactors are been adapted to industry in various forms as per the specific need of process.

This book provides the hydrodynamic and mass transfer characteristics of a downflow gas-interacting slurry bubble column reactor. The main key topics that are described in the book are gas entrainment, phase fraction, frictional drag, phase mixing, bubble–phase interactions, and solid–wall mass transfer phenomena.

The author provides the reader with a very thorough account of the fundamental principles and their applications to engineering practice, including a survey of the recent developments of three-phase flow in upflow gas-interacting slurry reactor and DGISR. The book will be useful in chemical and biochemical industries and academic and industrial research and development sectors, furthering understanding of the multiphase behavior in slurry bubble column reactor and aiding in the design and installation of the same in the industry for practical applications

PREFACE

Slurry bubble column reactors are intensively used as a multiphase reactors in chemical, biochemical, and petrochemical industries for carrying out reactions and for mass transfer operations in which a gas, made up of one or several reactive components, comes into contact or reacts with a liquid. The reactor is a simple vertical cylindrical vessel with a gas distributor at the bottom or top. The liquid phase may be supplied in batch or it may be moved with or against the flow of the gas.

Recently, various modified slurry bubble columns are gaining importance as a simple and inexpensive of means of achieving multiphase process yield due to their several advantages. Gas distribution plays an important role to influence the gas holdup, interfacial area and the degree of mass transfer. A plunging liquid jet with ejector system is an efficient alternative in this context. It provides high momentum to enhance the liquid mixing and gas distribution. Also, it provides efficient gas entrainment. Gas entrainment by plunging liquid jets is frequently met in practice. Gas entrainment is desirable to achieve gas absorption coupled with good mixing in some multiphase reactors. In particular, due to favorable energy requirements, a jet-induced gas distributor has potential applications in many chemical, fermentation, and wastewater treatment processes. Recently, downflow gas-interacting slurry bubble column reactors with jet-induced gas distributors for improved mixing have been recommended for many industrial processes such as aerobic fermentation, wastewater treatment, desorption and scrubbing absorption and so forth.

From the literature, it is found that no studies on hydrodynamics in the downflow gas-interacting slurry reactor (DGISR) are available, though it has several industrial applications. There is scope to study the industrial application of DGISR. The book provides the hydrodynamic and mass transfer characteristics of downflow gas-interacting slurry bubble column reactor.

The interest in the applications of gas-interacting downflow slurry bubble column reactor (GIDSBCR) for the process intensifications of chemical and biochemical industries is due to the following advantages: bubbles are finer and more uniform in size, coalescence of bubbles are negligible, homogenization of the two phases in the whole column is possible, a large amount of liquid can be contacted with a small amount of dispersed gas efficiently, and there is higher residence time of the gas bubbles, lower power consumption, almost complete gas utilization, higher overall mass transfer coefficient and tolerance to particulates, and therefore viable for slurry chemical reactions. Slurry reactors are been adapted in industry in various forms as per specific need of process. In this book, general features of the slurry reactor and its applications, advantage and disadvantage and their hydrodynamic behaviors are discussed. The gas-interacting downflow slurry reactor is one of the advanced reactors where residence time of the gas phase can be increased. The process description of the gas-interacting downflow slurry reactor on which the book is mainly focused is also described. The way of gas distribution plays an important role in gas-interacting three-phase slurry reactor. In case of the upflow, the distribution of gas is made by different types of spargers. However, in the case of downflow slurry reactor, the distribution is quite complex. It depends on the mechanism of the entrainment of gas to disperse in the reactor, which interacts with the liquid and solid to yield the process output. In this book, the mechanism of gas entrainment in DGISR, the conditions for minimum gas entrainment, and the energy efficiency of gas entrainment in the slurry reactor are enunciated. The three-phase pressure drop is an essential element for the design of process equipment, which is the subject of numerous experimental and numerical studies. To design and control such equipment efficiently, the knowledge of pressure drop of the system in the equipment is required. A chapter analyzes the studies of frictional pressure drop of three-phase flow in the DGISR. Bubble size distribution is a crucial factor to assess the interfacial area in mass transfer phenomena. It controls the efficiency of mass transfer in a multiphase reactor. The size distribution is related to the interfacial area of the phases where the diffusion of solute occurs on its surface, which is the backbone of the interfacial mass transfer in a reactor. In this book, the profile of bubble size distribution in the gas-interacting downflow slurry reactor is described. The interfacial area and the mass transfer characteristics are also described in the book.

—**Subrata Kumar Majumder**

ACKNOWLEDGMENTS

I would first like to acknowledge the efforts of our organization, the Indian Institute of Technology Guwahati, for providing an infrastructure for research and for writing the book. I would also like to acknowledge the constant support and encouragement of my wife. I am grateful to my former student Late Mekala Sivaiah for his contribution to some parts of the book.

I would like to express my gratitude to the reviewers for their constructive suggestions and comments to qualify the book. I am also grateful to the editorial and production departments of Apple Academic Press who were extremely cooperative in this endeavor.

—Subrata Kumar Majumder

CHAPTER 1

INTRODUCTION

CONTENTS

1.1   INTRODUCTION

A gas-interacting slurry reactor as a three-phase contactor is common in various chemical and biochemical processes. The slurry reactor are adapting in industry in various forms as per specific need of process. In this chapter, general features of the slurry reactor and its applications, advantage and disadvantage and their hydrodynamic behaviours are briefed. The gas-interacting downflow slurry reactor is one of the advanced reactors where contact time of the gas can be increased. The process description of the gas-interacting downflow slurry reactor on which the book is mainly focused is also described in this chapter.

1.2   THREE-PHASE REACTOR

The three-phase reactor is a multiphase contactor, where gas is distributed in a liquid in the presence of solid, which is used in various chemical and biochemical processes. The reactors can be designed either in for semi-batch or continuous-mode operation. The gas comes in contact with the liquid and diffuses into the liquid from the surface of the gas bubble. The diffused gas is then transferred from the liquid to the solid surface, where then the reaction takes place. They are very common in the chemical, petroleum, food, mining, pharmaceutical and semiconductor industries for the specific application, where gas is reacted with the liquid in the presence of catalyst particle.

1.2.1   TAXONOMY OF GAS–LIQUID–SOLID SLURRY REACTOR

Various chemical processes in gas–liquid–solid reactors are carried out based on the flow patterns of the slurry. It can be broadly split into concurrent and countercurrent flows based on the direction of the gas and slurry flows (Epstein, 1981). Taxonomy of the flow operation is shown in Figure 1.1.

Cocurrent flow: In cocurrent flow, contacting modes are upflow and downflow, which characterize different hydrodynamic characteristics between solid particles and the enclosing gas and liquid as shown in Figure 1.2.

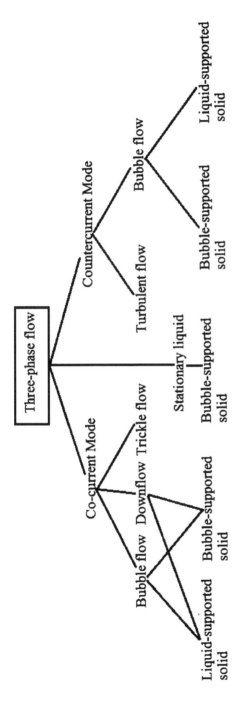

FIGURE 1.1   Taxonomy of three-phase flow.

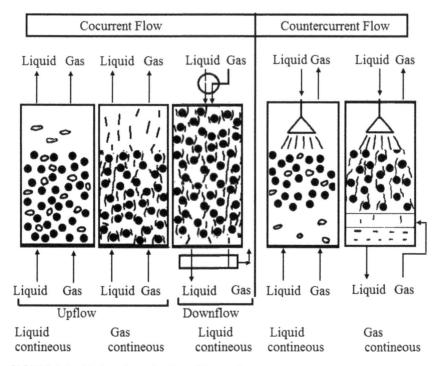

FIGURE 1.2 Modes of gas–liquid–solid operation.

Cocurrent upflow: The concurrent upflow modes are operated with two ways. One is called liquid-supported solid operation, while the other one is called bubble-supported solid operation. In case of liquid-supported solid operation, the liquid is in a continuous phase, where the solids are supported by the liquid with the dispersed gas bubbles. Epstein (1981) reported that the liquid-supported solids operation is governed at a minimum liquid velocity relative to solid terminal velocity. Different flow patterns such as coalesced bubble flow, dispersed bubble flow and slug flow are observed depending on the flow rate of the gas and liquid or slurry (Anderson and Quinn, 1970). In case of bubble-supported solids operation, the operation is characterized with the liquid velocity below the terminal velocity of solid or with the stationary liquid state in liquid batch. An increase in the gas flow rate in liquid-supported solid operation increases the formation of slug and decreases the liquid hold-up in the liquid–solid suspension. After an increase in the gas flow rate, the slurry forms segregated agglomerates and consequently bubble-supported operation prevails, in which the gas flows continuously and the liquid flows as a discrete phase either in as a flow of thin liquid films or as a liquid droplets (Mukherjee et al., 1974).

Cocurrent downflow: In this flow, the gas and liquid–solid mixture flow downward. The gas is distributed by specially designed distributor. The gas can be distributed in the column by liquid jet through an ejector (Sivaiah and Majumder, 2012). The jet is used to plunge on the fluid surface in the slurry medium in a column. During plunging, the gas is entrained by breaking the fluid surface and carries the gas downward as a dispersed phase of bubble. In the gas distribution zone, the bubbles are highly interacting with each other and continuously change their shape and size. The breakup of bubbles results in the formation of smaller bubbles and they are dragged downward by liquid momentum against their buoyancy. At high momentum (jet velocity > 15 m/s), the elongated bubbles are instantly formed, which may be referred as a slug flow condition. At this condition, the plunging zone increases by diminishing the bubbly zone (Majumder, 2016). In this regard, it is worth to include the invention of Shimodaira et al. (1981), which reports on biological treatment of waste water using a carrier floatable on water, and more particularly to a novel process for treatment of waste water utilizing a fluidized bed, which is formed by supplying waste water in a downflow operation. They claimed that the process is applicable to both anaerobic and aerobic biological treatment. The gas distribution without ejector system in case of downward flow may also result in annular, intermittent, separated, and dispersed flow patterns with only minor modifications (Crawford et al., 1986).

Countercurrent flow: This mode of operation is persuaded with liquid and gas, both as the continuous phases. The countercurrent gas-aided slurry reactor with continuous liquid phase is called inverse three-phase fluidization, whereas the reactor operated with continuous gas is called as a turbulent contact slurry bed reactor. In such cases, low-density solids relative to liquid is fluidized in a downflow contactor, by balancing the terminal velocity of the solid. The gas is flowed countercurrently to the liquid. Different flow regimes such as fixed bed with dispersed bubble, bubbling, transition, and slugging regime are observed in this operation (Fan et al., 1982a,b). In case of operation with continuous gas phase, a bed of low-density particles moves upward by the upward flow of gas continuously. At turbulent condition, the vigorous movement of particles results in tremendous gas–liquid contacting in the slurry column reactor. The taxonomy of the gas-interacting slurry reactor relates to the size of the reactor, methods of phase distribution and the reactor's internals.

1.2.2 GAS-INTERACTING SLURRY REACTOR

A gas-interacting slurry reactor is defined as a three-phase bubble column reactor utilizing the catalyst as a fine solids suspension in a liquid. It is used to carry out reactions between gas made up of one or several reactive components and liquid in the presence of catalyst particles. The reactors are a simple cylindrical vessel with gas distributor at bottom or top. The liquid phase may be supplied in batch or it may move with or against the flow of the gas. If the reactions are highly exothermic, an internal heat exchanger is to be used in the reaction zone. Except for the presence of solids, this type of slurry reactor is identical to the bubble column reactor commonly used for gas–liquid contacting accompanied by the chemical reaction. In the condition where the gas solubility is low (liquid-phase mass transfer is important) and a large liquid hold-up is required, this type of reactor is ideal. In contrast to physical mass transfer operations, counter flow offers no significant advantages as the reaction itself ensures a sufficient concentration drop during the material exchange. The top of the bubble column is often widened to facilitate gas separation. The bubble column reactor is characterized by lack of any mechanical means of agitation; hence, gas is distributed more evenly in the liquid phase. The gas-aided slurry reactors offer several advantages in comparison to other kinds of multiphase reactors. They are subjected to an increasing use in industrial applications in various chemical and biochemical processes such as coal liquefactions, absorption, desorption and hydrocarbon synthesis. When compared to other conventional multiphase reactors (where solids are not moving with liquid or gas), the gas-interacting slurry reactors have several advantages through higher rate of heat and mass transfer, high selectivity and conversion per pass, less maintenance cost due to their easy procurement, the absence of moving parts, higher gas hold-up and producing more interfacial area and higher dispersion efficiency. According to industrial practice, such reactors are modified in suitable forms. The different types of modified gas-interacting slurry reactors are shown in Table 1.1.

TABLE 1.1 Types of Gas-Interacting Slurry Reactors.

Type based on	Type	Description
Flow direction	Upflow	Liquid and gas flows concurrently in upward directions.
	Downflow	Flow of gas is downward with the upward and downward liquid flow.
Directional liquid circulation	External circulation	Liquid removal line is in between the dispersed phase where the elimination of radial transfer of solute over cross-sectional area takes place.
	Internal circulation	In this system, liquid removal line is adjacent to dispersed phase.
	Cascade	A bubble column cascade reactor comprising a vertical column and a plurality of equidistantly spaced, horizontally mounted, uniformly perforated plates therein.
Mechanical modifications	With static mixers	Static mixers are motionless mixtures; they are found to be effective as they increase the rates of mass transfer.
	Multi layered	In a multi-shaft bubble column, the presence of shafts prevents the coalescence of the bubbles. The shafts provide directed vertical paths for the bubbles preventing lateral movement.

Many gas-interacting slurry reactors offer directional fluid circulation. The simplest case uses the major effect resulting from bubble entrainment and the difference in density between the dispersed phase and the continuous phases. The large loop reactors are also designed with either an internal or external liquid removal line, depending on the gas removal arrangement if required. Hines (1978) developed a shaft reactor based on a circulation process which is a highly significant development in the column reactor. The advantage of this arrangement is that large amount of gas reach the lower parts of the shaft (as shown in the Fig. 1.1). The column with external circulation systems is similar to loop reactors. As shown in Figure 1.1a, the shaft at the centre eliminates transverse transfer over whole cross-sectional area resulting with a homogeneous flow pattern. The dispersed phase is in centre of the column and slurry circulation occurs through annulus region. With internal circulation system as shown in Figure 1.3b, dispersed phase exists in annulus region, whereas slurry circulation occurs through the centre of the column. These types of reactors provide homogenous flow zone and high rate of circulation. The inserted loop stabilizes the bulk circulation. In a multi-shaft column as shown in Figure 1.4a, the presence of shaft prevents the coalescence of the bubbles. The shaft provides directed vertical paths for the bubbles preventing lateral movement and bulk circulation. The shaft could also act as a cooling device to control the temperature of mixture inside the column. The column with static mixers consists of motionless mixers (Fig. 1.4b). They are found to be effective as they increase the rates of mass transfer and also helps in bubble flow to remain homogeneous. The multi-stage bubble columns as shown in Figure 1.4c change the retention time of the slurry phase. A cascade gas-interacting slurry reactor as shown in Figure 1.4d is a vertical column with equidistantly spaced, horizontally mounted and uniformly perforated plates fitted inside it. The intermixing and the non-uniform bubble size distribution can be changed by the fitting of different trays. The packed bubble columns as shown in Figure 1.4e has high hold-up and better cross-sectional distribution of the liquid phase, which make them superior to trickle bed reactors.

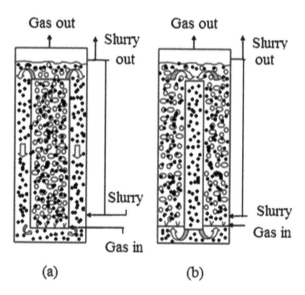

FIGURE 1.3 (a) Bubble column with external circulation, (b) bubble column with internal circulation (Majumder, 2016)

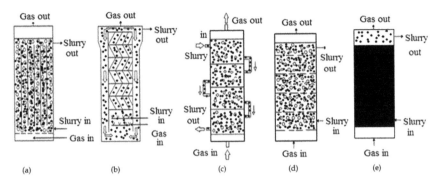

FIGURE 1.4   (a) Multi-shaft bubble column, (b) bubble column with static mixture, (c) multi-stage bubble column, (d) cascade bubble column, (e) packed bubble columns (Majumder, 2016).

The gas-interacting slurry reactors with forced circulation use the liquid jet as a means of gas distribution and the energy involved in this process generates the circulation. In the case of the slurry jet loop reactor, the gas is finely atomized in the shear field of the slurry jet and the reactor contents are efficiently circulated by the means of conduit tube. If the gas is forced in through nozzles and a momentum exchange unit incorporated, then this is known as blast nozzle reactor. In the free jet configuration the gas is efficiently distributed through two-component nozzles and slurry circulation is strongest in the lower section of the reactor. For intensifications, the gas–liquid–solid process. The ejector-induced fluid mixing devices are also good alternative for achieving efficient dispersion of one fluid into other (Sivaiah et al., 2012; Sivaiah and Majumder, 2012a,b, 2013a-c). Liquid jet through ejector system is considered to be useful for gas–liquid–solid contactors as it generates fine bubbles by impinging the slurry jet on the slurry surface. Another advantage of the slurry jet is that it entrains certain amount of gas into the pool of solid–liquid slurry. Liquid jet reactor is similar to a gas-dispersed stirred tank reactor with ejector for both mixing and gas entrainment (Burgess and Molloy, 1973; Ogawa et al., 1986; Sivaiah and Majumder, 2013b). Bin (1993) reported the review on the several aspects and phenomena of gas entrainment in a multiphase reactor. He also reported the practical applications of the plunging jet gas-interacting slurry reactor in waste water treatment, fermentation and mineral beneficiation. Some important studies on slurry reactor are compiled in Table 1.2.

TABLE 1.2 Literature Database for Parametric Studies of Gas-Aided Slurry Column Reactor.

Author	System studied	Range of operating variables	Gas distributor	Column geometry	Parameters studies
Kölbel et al. (1971)	**Gas:** H_2 and ethylene; **Liquid:** C_{13-18} mixture; **Solid:** catalyst particles	P: 0.1–0.558 MPa; T: 298–353 K; u_{sg}: 0–0.17 m/s; u_{sl}: 0 m/s; C_v: 3%	Porous plate ($d_0 = 10$ μm)	H_c: 4 m; d_c: 0.0418 m	Gas hold-up and axial dispersion coefficient
Deckwer et al. (1980)	**Gas:** N_2; **Liquid:** molten paraffin; **Solid:** alumina powder (5 μm)	P: 0.4–1.1 MPa T: 416–533 K; u_{sg}: 0–0.035 m/s; u_{sl}: 0 m/s; C_v: 0–16%	Perforated plate $d_0 = 75$ μm	H_c: Unknown; d_c:0.041, 0.1 m	Gas hold-up
Sangnimnuan et al. (1984)	**Gas:** air; **Liquid:** tetralin; **Solid:** coal	P: 4.5–15 MPa; T: 437–657 K; u_{sg}: 0.012–0.02 m/s; u_{sl}: 0.001–0.003 m/s; C_v: unknown	Unknown		Gas hold-up and axial dispersion coefficient
Fukuma et al. (1987)	**Gas:** air; **Liquid:** water and aqueous solution of glycerol; **Solid:** glass beads (56–460 μm)	P: 0.1 MPa; T: 293.05 K; u_{sg}: 0.01–0.08 m/s; u_{sl}: 0.001–0.05 m/s; C_v: 0–50%	Multi-nozzle (8 tubes, $d_0 = 2.6$ mm)	H_c: 1.2–3.2 m; d_c: 0.15 m	Bubble size
Öztürk et al. (1987)	**Gas:** air; **Liquid:** water, ligroin, tetralin and aqueous solution Na$_2$SO$_4$ (0.8 M); **Solid:** PE, PVC, activated carbon and Kieselguhr, Al$_2$O$_3$	P: 0.1 MPa; T: 293 K; u_{sg}: 0–0.08 m/s; u_{sl}: 0 m/s; C_v: 0%	Two single orifice tubes ($d_0 = 3$ and 0.9 mm)	H_c: 0.85 m; d_c: 0.095 m	Gas hold-up and mass transfer coefficient

TABLE 1.2 (Continued)

Author	System studied	Range of operating variables	Gas distributor	Column geometry	Parameters studies
Clark (1990)	**Gas:** N_2 and H_2; **Liquid:** water and methanol; **Solid:** glass powder (45–63 μm)	P: 2.5–10 MPa T: 293–453 K; u_{sl}: 0 m/s; 0–0.6 m/s; u_{sg}: C_v: 0–10%	Porous plate $d_0 = 60$ μm	$H_C = 1.6$ m; $d_C = 0.075$ m	Gas hold-up
Reilly et al. (1994)	**Gas:** air, Ar and He; **Liquid:** water, Varsol, Varsol + antifoam, trichloroethylene; **Solid:** glass beads (71–745 μm)	P: 0.1 MPa; T: 283–323 K; u_{sg}: 0–0.35 m/s; u_{sl}: 0 m/s; C_v: 0–25%	Perforated plate (293 orifices, $d_0 = 1.5$ mm), single orifice pipe ($d_0 = 25.4$ mm), spider sparger (6 orifices, $d_0 = 13.4$ mm)	H_C: 5 m; d_C: 0.30 m	Gas hold-up
Dewes et al. (1995)	**Gas:** air, **Liquid:** water, 0.8 M sodium sulphate solution; **Solid:** glass beads	P: 0.1–0.8 MPa T: 298 K; u_{sg}: 0.01–0.08 m/s; u_{sl}: 0 m/s; C_v: 0–2%	Perforated plate ($d_0 = 1$ mm, 7 holes)	H_C:1.37 m; d_C: 0.115 m	Gas hold-up, mass transfer coefficient and interfacial area
De Swart and Krishna (1995)	**Gas:** air; **Liquid:** mineral oil, ethanol, octanol, water and aqueous solutions of NaOH; **Solid:** glass beads (40 μm)	P: 0.1 MPa T: 298 K; u_{sg}: 0–0.5 m/s; u_{sl}: 0 m/s; C_v: 0–20%	Porous plate ($d_0 = 0.2$ mm), Bronze sintered plate ($d_0 = 50$ μm)	H_C: 3–4.5 m; d_C: 0.05–0.174 m	Gas hold-up
Dewes and Schumpe (1997)	**Gas:** He, N_2, air and SF_6; **Liquid:** water and 0.8 M sodium sulphate solution; **Solid:** xanthan gum, kieselguhr (22 μm), alumina (7 μm)	P: 0.1–1 MPa T: 298 K; u_{sg}: 0.01–0.08 m/s; u_{sl}: 0 m/s; C_v: 0–18%	Perforated plate ($d_0 = 1$ mm, 7 holes)	H_C 1.37 m; d_C 0.115 m	Gas hold-up and mass transfer coefficient

TABLE 1.2 *(Continued)*

Author	System studied	Range of operating variables	Gas distributor	Column geometry	Parameters studies
Fan et al. (1999)	**Gas:** N_2; **Liquid:** Paratherm NF; **Solid:** glass beads	P: 0.1–15.6 MPa; T: 299.5–360.5 K; u_{sg}: 0.05–0.69 m/s; u_{sl}: 0 m/s; C_v: 0%	Perforated plate, bubble cap, and sparger	H_C: >3.05 m; d_C: 0.1–0.61 m	Gas hold-up and bubble size
Luo et al. (1999)	**Gas:** N_2; **Liquid:** Paratherm NF; **Solid:** alumina (100 μm)	P: 0.1–5.6 MPa; T: 301–351 K; u_{sg}: 0–0.045 m/s; u_{sl}: 0 m/s; C_v: 0–19.1%	Perforated plate (120 holes, $d_0 = 1.5$ mm)	H_C: 1.37 m; d_C: 0.102 m	Gas hold-up and bubble size
Gandhi et al. (1999)	**Gas:** air; **Liquid:** water; **Solid:** glass beads (35 μm)	P: 0.1 MPa; T: 293.5 K; u_{sg}: 0.05–0.28 m/s; u_{sl}: 0.001–0.05 m/s; C_v: 0–40%	Four armed sparger (20 orifices, $d_0 = 1.5$ mm)	H_C:2.5 m; d_C: 0.15 m	Gas hold-up
Onozaki et al. (2000)	**Gas:** H_2; **Liquid:** oil slurry; **Solid:** coal + slurry	P: 16.6–16.8 MPa; T: 313–733 K; u_{sg}: 0.01–0.09 m/s; u_{sl}: 0.002–0.004 m/s; C_v: 0–51.5%	Unknown	H_C: 11 m; d_C: 1 m	Gas hold-up
Yang et al. (2001)	**Gas:** $CO–H_2–N_2$; **Liquid:** liquid paraffin; **Solid:** silica gel powder (134 μm)	P: 1–5 MPa; T: 293–523 K; u_{sg}: 0–0.025 m/s; u_{sl}: 0 m/s; C_v: 0–20%	Unknown	H_C: 0.48 m; d_C: 0.037 m	Mass transfer coefficient and interfacial area

TABLE 1.2 *(Continued)*

Author	System studied	Range of operating variables	Gas distributor	Column geometry	Parameters studies
Ishibashi et al. (2001)	**Gas:** H_2; **Liquid:** oil slurry; **Solid:** coal + catalyst	P: 16.8–18.6 MPa; T: 505–738 K; u_{sg}: 0.01–0.09 m/s; u_{sl}: 0.002–0.004 m/s; C_V: Unknown	Unknown	H_C: 11 m; d_C: 1 m	Gas hold-up
Therning and Rasmunson (2001)	**Gas:** air; **Liquid:** water; **Solid:** plastic ball rings packing	P: 0.1–0.66 MPa; T: 298 K; u_{sg}: 0.03–0.17 m/s; u_{sl}: 0 cm/s; C_V: 0%	Perforated plate (d_0 = 1.5 mm), 0.2% open area	H_C: 3.2 m; d_C: 0.154–0.2 m	Gas hold-up and axial dispersion coefficient
Behkish et al. (2002)	**Gas:** H_2, CO, N_2 and CH_4 **Liquid:** Isopar-M, hexanes mixtures **Solid:** iron oxides and glass beads	P: 0.17–0.8 MPa; T: 298 K; u_{sg}: 0.08–0.2 m/s; U_L: 0 m/s; C_V: 0–36%	Unknown	H_C: 2.8 m; d_C: 0.316 m	Gas hold-up, bubble size and mass transfer coefficient
Kluytmans et al. (2003)	Gas: air, O_2, N2; liquid: water and sodium gluconate aqueous solution; Solid: carbon (30 µm)	P: 0.1 MPa; T: 298 K; u_{sg}: 0–0.4 m/s; u_{sl}: 0 m/s; C_V: unknown %	Unknown	H_C: 2.0 m; d_C: 0.3 m	Gas hold-up, bubble size and mass transfer coefficient, interfacial area
Jin et al. (2004)	**Gas:** air; **Liquid:** water; **Solid:** quartz sand	P: 1–3 MPa; T: 298–473 K; u_{sg}: 0.03–0.1 m/s; u_{sl}: 0 m/s; C_V: 0–20%	Four nozzles (d_0 = 8 mm)	H_C: 0.4–0.6 m; d_C: 0.06–0.1 m	Mass transfer coefficient

TABLE 1.2 *(Continued)*

Author	System studied	Range of operating variables	Gas distributor	Column geometry	Parameters studies
Vandu et al. (2004)	**Gas:** air; **Liquid:** paraffin oil; **Solid:** alumina based—catalyst	P: 0.1 MPa; T: 298 K; u_{sg}: 0–0.4 m/s; u_{sl}: 0 m/s; C_v: 0–25%	Perforated plate (199 holes, $d_0 = 0.5$ mm)	H_C: 1.34–1.36 m; d_c: 0.1 m	Gas hold-up and mass transfer coefficient
Kantarci et al. (2005)	**Gas:** air; **Liquid:** water; **Solid:** yeast cells (10 μm) and bacteria cells (0.2–0.7 μm)	P: 0.1 MPa; T: 298 K; u_{sg}: 0.02–0.25 m/s; u_{sl}: 0 m/s; C_v: 0%	Spider-type (6 arms, 24 holes, $d_0 = 2$ mm)	H_C: 0.6 m; d_c: 0.17 m	Gas hold-up and bubble size distribution
Cui (2005)	**Gas:** air; **Liquid:** water, Norpar 15; **Solid:** glass beads (120 μm) and acetate (500–2000 μm)	P: 0.1–1.5 MPa T: 298 K; u_{sg}: 0–0.201 m/s; u_{sl}: 0 m/s; C_v: 0–4%	Perforated plate ($d_0 = 1.5$ mm, 120 holes), single nozzle ($d_0 = 6$ mm)	H_C: 0.15–1.37 m; d_c: 0.0508–1016 m	Gas hold-up
Behkish et al. (2007)	**Gas:** N_2–He; **Liquid:** Isopar-M; **Solid:** alumina powder	P: 0.67–3 MPa T: 298–473 K; u_{sg}: 0.07–0.39 m/s; u_{sl}: 0 m/s; C_v: 0–20%	Spider-type (6 legs, 108 orifices, $d_0 = 5$ mm)	H_C: 3 m; d_C: 0.29 m	Gas hold-up and bubble size
Hashemi et al. (2009)	**Gas:** N_2/O_2 mixture; **Liquid:** water; **Solid:** ion-exchange resin (84.8 μm)	P: 0.1–4.0 MPa; T: 277–295 K; u_{sg}: 0–0.2 m/s; u_{sl}: 0 m/s; C_v: 0–10%	Perforated plate (34 holes, $d_0 = 3.175$ mm)	H_C: 0.6–0.7 m; d_c: 0.1	Gas hold-up and mass transfer coefficient

TABLE 1.2 *(Continued)*

Author	System studied	Range of operating variables	Gas distributor	Column geometry	Parameters studies
Chilekar et al. (2010)	**Gas:** N_2 and air; **Liquid:** water, aqueous solution of sodium gluconate and Isopar-M; **Solid:** carbon (30 μm) and silica (44 μm)	P: 0.1–1.3 MPa T: 292–296 K; u_{sg}: 0–0.4 m/s; u_{sl}: 0 m/s; C_v: 0–3%	Perforated plate ($d_0 = 0.5$ mm, 200 holes)	$H_c = 1.4$ m; $d_c = 0.15$ m	Gas hold-up, bubble size and mass transfer coefficient
Mena et al. (2011)	**Gas:** air; **Liquid:** water; **Solid:** EPS (1100, 770 and 591 μm) and glass beads (9.6 μm)	P: 0.1 MPa; T: 298 K; u_{sg}: 0–0.0027 m/s; u_{sl}: 0 m/s; C_v: 0–30%	Multi-nozzle sparger (13 needles, $d_0 = 0.3$ mm)	H_c: 0.2 m; d_c: 0.084 m	Mass transfer coefficient and interfacial area
Kumar et al. (2012)	**Gas:** air; **Liquid:** water; **Solid:** glass beads (35 μm)	P: 0.1 MPa; T: 298 K; u_{sg}: 0.01–0.1628 m/s; u_{sl}: 0–0.1226 m/s; C_v: 0–9%	Perforated plate (d_0 unknown)	H_c: 1–1.8 m; d_c: 0.15 m	Gas hold-up
Sivaiah et al. (2012a, b)	**Gas:** air; **Liquid:** water and electrolyte solution; **Solid:** zinc oxide, aluminium oxide and kieselguhr	P: atmospheric T: 295 K u_{sg}: 0–0.034 m/s u_{sl}: 0–0.13 m/s Cv: 0.1–1.0	Ejector	Hc: 1.6 m dc: 0.05 m	Gas hold-up and entrainment characteristics
Sivaiah and Majumder (2012)	**Gas:** air; **Liquid:** water and electrolyte solution; **Solid:** zinc oxide, aluminium oxide and kieselguhr	P: atmospheric T: 295 K u_{sg}: 0–0.034 m/s u_{sl}: 0–0.13 m/s C_v: 0.1–1.0	Ejector	H_c: 1.6 m d_c: 0.05 m	Gas hold-up and frictional pressure drop

TABLE 1.2 *(Continued)*

Author	System studied	Range of operating variables	Gas distributor	Column geometry	Parameters studied
Sivaiah and Majumder (2013a, b, c)	**Gas:** air; **Liquid:** water and electrolyte solution; **Solid:** zinc oxide, aluminium oxide and kieselgur	P: atmospheric T: 295 K u_{sg}: 0–0.034 m/s u_{sl}: 0–0.13 m/s C_v: 0.1–1.0	Ejector	H_C: 1.6 m d_c: 0.05 m	Gas hold-up, mass transfer coefficient, Axial mixing, frictional resistance
Jin et al. (2014)	**Gas:** H_2, CO and CO_2; **Liquid:** paraffin; **Solid:** quartz sand (150–200 μm)	P: 1–3 MPa; T: 298–473 K; u_{sg}: 0.03–0.1 m/s; u_{sl}: 0 m/s; C_v: 0–20%	Perforated plate ($d_0 = 8$ mm, 2.56% open area)	H_C: 0.4–0.6 m; d_c: 0.10 m	Mass transfer coefficient and interfacial area
Li and Zhong (2015)	**Gas:** air; **Liquid:** water **Solid:** glass powder	P: atmospheric T: ambient u_{sg}: 0.25, 0.1, 0.036, 0.089, 0.16, 0.22, 0.33 m/s u_{sl}: 0 m/s Cv: 3–30%	Not known	Hc: 2.5, 2.0, 0.8; d_c: 0.15, 0.20, 0.1 × 0.1 m (L × B)	Hydrodynamics of gas–liquid–solid three-phase, gas hold-up
Li et al., 2016	**Gas:** air; **Liquid:** deionized water, glycerol–water and deionized water with surfactant; **Solid:** glass beads and silica gel	P: atmospheric T: ambient u_{sg}: 8.8×10^{-6} to 4.4×10^{-4} m/s u_{sl}: 0–7.5 mm/s C_v: 3–30%	Porous sponge supported a wire fine-mesh with mesh size of 47 μm	H_C: 50 mm d_c: 2–10 mm dp: 80–200 μm	Pressure drop, Bubble characteristics

TABLE 1.2 *(Continued)*

Author	System studied	Range of operating variables	Gas distributor	Column geometry	Parameters studies
Nedeltchev (2017)	**Gas:** air, nitrogen, hydrogen and helium; Liquid: decalin, 1,2-dichloroethane, 1,4-dioxane, ethanol (99%), nitrobenzene, 2-propanol, ethyl-englycol, tetralin, xylene and tap water, mixture of water–glycol (22.4, 60.0 and 80.0%) and toluene–ethanol (94.3 and 97.2%) Solid: kieselgur (diatomaceous earth) and activated carbon	P: not known T: not known u_{sg} up to 0.08 m/s; $u_{sl} = 0$ m/s	Single tube ($\varnothing\, 0.9 \times 10^{-3}$, $3.0 \times 10^{-3\,m}$)	H_C: not known d_C: 0.095 m	Bubble size, mass transfer coefficient

1.2.3   DOWNFLOW GAS-INTERACTING SLURRY REACTOR (DGISR) AND ITS OPERATION

In the downflow gas-interacting slurry reactor (DGISR), the gas and slurry are entered from upper part of the reactor as shown in Figure 1.5 by specially designed gas and liquid distributor (for gas ejector system, and for liquid, a nozzle). The slurry is introduced as slurry jet which plunges into the free surface of the phase mixture in a contactor. The nozzle induces gas from the source and entrains into the pool of slurry in the contactor by jet kinetic energy.

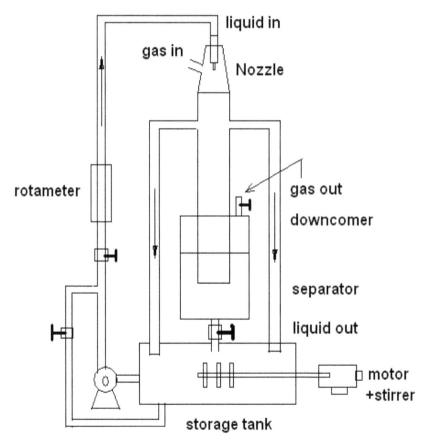

FIGURE 1.5   Schematic diagram of downflow gas-interacting slurry reactor (DGISR).

The rate of entrainment may depend on the slurry jet velocity and the three-phase gas–liquid–solid mixing height inside the contactor or column. The zone where liquid jet impinges on the free surface of the gas–liquid–solid mixture is called mixing zone. In the mixing zone, the gas is entrained and is broken into fine bubbles and dispersed in the column. After that, the fine bubbles are transported by the slurry downward through the column. A fraction of the entrained gas as a dispersed phase of bubbles moves back into the surface and is dissolved there, whereas other fraction moves downward. This depends on the buoyancy of the gas bubble. The buoyancy is directly related to the bubble size. Finer bubble have lower buoyancy, whereas coarser have higher buoyancy due to which coarser bubble moves upward relative to the finer bubble. The gas as the dispersed finer bubbles is transported by the downward liquid momentum and gets separated from the slurry in a separator. The gas-interacting downflow slurry reactor can be procured with an ejector assembly for gas induction, an extended pipeline contactor for gas–liquid (Ohkawa et al., 1986; Evans et al., 1992; Lu et al., 1994; Majumder et al., 2005; Sivaiah and Majumder, 2012; Sivaiah and Majumder, 2013a,b,c; Majumder, 2016) and solid mixing and downward transportation with well mixing, a gas–slurry separator to separate the gas from the slurry and settling of solid and other accessories as shown in Figure 1.5.

The ejector assembly is used to suck the maximum gas without any compressor at an optimum suction pressure. The lower end of the contactor is prolonged sufficiently inside the separator to allow uniform movement of the phases by developing uniform flow in the contactor and tranquil separation of the gas from the slurry. The ejector assembly, the nozzle and the contactor are flawlessly aligned vertically to obtain an axially symmetric jet. When the jet plunges into the pool of the slurry in the column, the gas is entrained as the bubbles. The level of the liquid–solid slurry inside the separator is to be maintained at a particular height by adjusting the valves to balance the hydrostatic pressure in the column and steady operation. The design, modelling and scale-up of the reactor require accurate knowledge of the kinetics, hydrodynamics and transport design parameters.

1.3   APPLICATION OF A GAS-INTERACTING SLURRY REACTOR

The gas-interacting slurry reactors are widely used in chemical, petrochemical and biological industries, where heterogeneous gas–liquid–solid reactions are to be performed. The reactors are typical representative of industrial equipment, where gas phase and liquid phase are contacted, usually to promote mass transfer across the interface. The multiphase reaction is performed in slurry reactors in a batch-wise manner. The application of the gas-interacting slurry reactors depends upon the method of addition of the liquid reaction mixture. The role of the hydrodynamics and transport phenomena is accountable for modeling, design and scale-up of the systems (Li et al., 2003; Baten et al., 2003; Strasser and Wonders, 2008). The gas-interacting slurry reactors have a long history of commercial use in specific applications. Some of these are briefed as follows.

1.4   SOME OTHER IMPORTANT THREE-PHASE APPLICATIONS

Some important three-phase flow applications in gas-interacting slurry reactor are mentioned as follows (Dudukovi´c et al., 2002):

- Desulphurization for the production of low-sulphur fuel oils (Benzaazoua et al., 2017)
- Hydrodesulphurization of atmospheric gas oil (El Sayed et al., 2017)
- Catalytic dewaxing (Köhler 2007; Belinskaya et al., 2017)
- Sweetening of oil (Ganguly et al., 2012)
- Slurry phase Hydrocracking for production of high-quality fuels (Nguyen et al., 2018)
- Hydrodenitrification (Chen et al., 2018)
- Cracking for the production of naphtha (Usman et al., 2017)
- Lubricating oil processing (Pereira et al., 2017)
- Synthesis from syngas (Zhang et al., 2014; Wang et al., 2016)
- Hydrogenation (Heck and Smith, 1970; Hichri et al., 1991; Gavroy et al.,1995; Alini et al., 2003; Marwana and Winterbottom, 2004; Bergault et al., 1999; Mathew et al., 1999; Chaudhari et al., 2003; Crespo-Quesada et al., 2009; Grasemann et al., 2010; Liedtke et al., 2013; Saeidi et al., 2014)
- Synthesis of butynediol from acetylene and aqueous formaldehyde (Yang et al., 2014)
- Hydration of propene (Marcì et al., 2017)
- Oxidation (Iliuta and Larachi, 2001; Kolaczkowski et al., 1999; Luck,1996; Pollington et al., 2009)

- Catalytic ozonation (Li et al., 2018)
- Oxidation cumene (Hattori et al., 1970), oxidation of SO_2; Oxidation of glucose, poly (α-olefin) and so forth.
- Absorption of gas in the presence of solid particles (Jana and Bhaskarwar, 2011)
- Desorption (Du et al., 2018; Spiandore et al., 2018)
- Fischer—Tropsch Synthesis (Fischer and Tropsch, 1926; Ledakowicz et al., 1985; Eliason and Bartholomew, 1999; Dry 2002; Anfray et al., 2007; Brady and Pettit, 1981; Chang et al., 2007; Chang et al., 2007)
- Methanation (Bian et al., 2015; da Silva et al., 2012; Gao et al., 2013, 2017; Krämer et al., 2009; Wang et al., 2016)
- Acid Rock Drainage Treatment (Xu et al., 2010, 2013)

1.5   ADVANTAGES OF A GAS-INTERACTING SLURRY REACTOR

When compared to other conventional three-phase reactors (e.g. trickle bed reactors and packed bed reactors), the gas-interacting slurry reactor reactors offers several advantages, which are enlisted below (Shah et al., 1982)

- Higher rate of heat and mass transfer
- High selectivity and conversion per pass
- The maintenance cost and requirements are very less due their easy procurement
- Economic energy consumption as the mixing is induced only by the gas aeration
- Higher gas hold-up controlling and interfacial area production
- Produces finer and uniform bubble
- Higher dispersion efficiency by utilizing the jet energy
- Higher residence time of the gas bubbles by balancing buoyancy force
- Constant overall catalytic activity maintained easily by the addition of small amount of catalyst
- Large heat capacity of reactor acts as a safety feature against explosions

1.6   CHALLENGES OF A GAS-INTERACTING SLURRY REACTOR

Although large number of advantages, several challenges, inherent to the gas-interacting slurry reactor reactors, however, have to be considered. This can be minimized by proper designing. The challenges are:

- A severe degree of backmixing may be in the liquid phase due to the low liquid flow rate for upward slurry reactor (Shah et al., 1978)
- Depositions of the solid particles at the bottom of the reactor
- High pressure drop with respect to packed columns which may require more energy to disperse the phase
- Uncertainties in design process
- Finding suitable liquid and catalyst may be difficult
- Size of the particle matters
- Higher ratio of liquid to catalyst than in other reactors

From the literature, it is found that there are no significant studies on hydrodynamics in the DGISR though it has a potential for several industrial applications by intensification of the retention time of the gas phase and reducing the backmixing of the liquid. Recently, downflow bubble column reactor with ejector type of gas distributor for improved mixing have been recommended for many industrial processes such as aerobic fermentation, waste water treatment, desorption and scrubbing absorption and so forth. (Deckwer, 1992; Majumder, 2016). There is a scope to study the industrial application of DGISR.

KEYWORDS

- **cocurrent downflow**
- **concurrent upflow modes**
- **countercurrent flow**
- **downflow gas-interacting slurry reactor (DGISR)**
- **gas-interacting slurry reactor**
- **gas–liquid–solid reactors**
- **three-phase reactor**

REFERENCES

Alini, S.; Bottino, A.; Capannelli, G.; Carbone, R.; Comite, A.; Vitulli, G. The Catalytic Hydrogenation of Adiponitrile to Hexamethylenediamine Over a Rhodium/Alumina Catalyst in a Three Phase Slurry Reactor, *J. Mol. Catal. A: Chem.* **2003**, *206*(1–2), 363–370.

Anderson, J. L.; Quinn, J. A. The Transition to Slug Flow in Bubble Columns. *Chem. Eng. Sci.* **1970**, *25*(2), 338–340.

Anfray, J.; Bremaud, M.; Fongarland, P.; Khodakov, A.; Jallais, S.; Schweich, D. Kinetic Study and Modeling of Fischer-Tropsch Reaction Over a Co/Al_2O_3 Catalyst in a Slurry Reactor. *Chem. Eng. Sci.* **2007**, *62*(18–20), 5353–5356.

Behkish, A.; Lemoine, R.; Sehabiague, L.; Oukaci, R.; Morsi, B. I. Gas Holdup and Bubble Size Behavior in a Large-Scale Slurry Bubble Column Reactor Operating with an Organic Liquid Under Elevated Pressures and Temperatures. *Chem. Eng. J.* **2007**, *128*, 69–84.

Behkish, A.; Men, Z.; Inga, J. R.; Morsi, B. I. Mass Transfer Characteristics in a Large-Scale Slurry Bubble Column Reactor with Organic Liquid Mixtures. *Chem. Eng. Sci.* **2002**, *57*, 3307–3324.

Belinskaya, N. S.; Frantsina, E. V.; Ivanchina, E. D. Mathematical Modelling of "Reactor—Stabilizer Column" System in Catalytic Dewaxing of Straight Run and Heavy Gasoils. *Chem. Eng. J.* **2017**, *329*, 283–294.

Benzaazoua, M.; Bouzahzah, H.; Taha, Y.; Kormos, L.; Kongolo, M. Integrated Environmental Management of Pyrrhotite Tailings at Raglan Mine: Part 1 Challenges of Desulphurization Process and Reactivity Prediction. *J. Cleaner Prod.* **2017**, *162*, 86–95.

Bergault, I.; Joly-Vuillemin, C.; Fouilloux, P.; Delmas, H. Modeling of Acetophenone Hydrogenation Over a Rh/C Catalyst in a Slurry Airlift Reactor. *Catal. Today* **1999**, *48*(1–4), 161–174.

Bian, L.; Zhang, L.; Xia, R.; Li, Z. Enhanced Low-Temperature CO_2 Methanation Activity on Plasma-Prepared Ni-Based Catalyst. *J. Nat. Gas Sci. Eng. 27 Part* **2015**, *2*, 1189–1194.

Bin, A. K. Gas Entrainment by Plunging Liquid Jet. *Chem. Eng. Sci.* **1993**, *48*, 3585–3630.

Brady, R. C.; Pettit, R. On the Mechanism of the Fischer-Tropsch Reaction. The Chain Propagation Step. *J. Am. Chem. Soc.* **1981**, *103*(5), 1287–1289.

Burgess, J. M.; Molloy, N. A. Gas Absorption in Plunging Jet Reactor. *Chem. Eng. Sci.* **1973**, *28*, 183–190.

Chang, J.; Bai, L.; Teng, B.; Zhang, R.; Yang, J.; Xu, Y.; Xiang, H.; Li, Y. Kinetic Modeling of Fischer—Tropsch Synthesis Over Fe-Cu-K-SiO2 Catalyst in Slurry Phase Reactor. *Chem. Eng. Sci.* **2007**, *62*(18–20), 4983–4991.

Chaudhari, R. V.; Rode, C. V.; Deshpande, R. M.; Jaganathan, R.; Leib, T. M.; Mills, P. L. Kinetics of Hydrogenation of Maleic Acid in a Batch Slurry Reactor Using a Bimetallic Ru–Re/C Catalyst. *Chem. Eng. Sci.* **2003**, *58*(3–6), 627–632.

Chen, B.; Xu, H.; Ni, M. Modelling of SOEC-FT Reactor: Pressure Effects on Methanation Process. *Appl. Energy* **2017**, *185*(Part 1), 814–824.

Chen, G.; Zhang, Z.; Zhang, Z.; Zhang, R. Redox-Active Reactions in Denitrification Provided by Biochars Pyrolyzed at Different Temperatures. *Sci. Total Environ.* **2018**, *615*, 1547–1556.

Chilekar, V. P.; Van der Schaaf, J.; Kuster, B. F. M.; Tinge, J. T.; Schouten, J. C. Influence of Elevated Pressure and Particle Lyophobicity on Hydrodynamics and Gas–Liquid Mass Transfer in Slurry Bubble Columns. *AIChE J.* **2010**, *56*, 584–596.

Clark, K. N. The Effect of High Pressure and Temperature on Phase Distributions in a Bubble Column. *Chem. Eng. Sci.* **1990**, *45*, 2301–2307.

Crawford, T. J.; Weinberger, C. B.; Weisman, J. Two-Phase Flow Patterns and Void Fractions in Downward Flow. Part II: Void Fractions and Transient Flow Patterns. *Int. J. Multiphase Flow* **1986**, *12*(2), 219–236.

Crespo-Quesada, M.; Grasemann, M.; Semagina, N.; Renken, A.; Kiwi-Minsker, L. Kinetics of the Solvent-Free Hydrogenation of 2-Methyl-3-Butyn-2-Ol Over a Structured Pd-Based Catalyst. *Catal. Today* **2009**, *147*(3–4), 247–254.

Cui, Z. Hydrodynamics in a Bubble Column at Elevated Pressures and Turbulence Energy Distribution in Bubbling Gas–Liquid and Gas–Liquid–Solid Flow Systems. Ph.D. Dissertation, the Ohio State University, Columbus, 2005.

da Silva, D. C. D.; Letichevsky, S.; Borges, L. E. P.; Appel, L. G., The Ni/ZrO$_2$ Catalyst and the Methanation of CO and CO$_2$. *Int. J. Hydrogen Energy* **2012**, *37*(11), 8923–8928.

De Swart, J. W. A.; Krishna, R. Influence of Particles Concentration on the Hydrodynamics of Bubble Column Slurry Reactors. *Chem. Eng. Res. Design* **1995**, *73*, 308–313.

Deckwer, W.-D.; Kokuun, R.; Sanders, E.; Ledakowicz, S. Kinetic Studies of Fischer-Tropsch Synthesis on Suspended Iron/Potassium Catalyst—Rate Inhibition by Carbon Dioxide and Water. *Ind. Eng. Chem. Process Des. Dev.* **1986**, *25*(3), 643–649.

Deckwer, W.-D. *Bubble Column Reactors*. John Wiley and Sons Ltd.: Chichester, England, UK, 1992, Chapter 1.

Dewes, I.; Kueksal, A.; Schumpe, A Gas Density Effect on Mass Transfer in Three-Phase Sparged Reactors. *Chem. Eng. Res. Design* **1995**, *73*, 697–700.

Dewes, I.; Schumpe, A Gas Density Effect on Mass Transfer in the Slurry Bubble Column. *Chem. Eng. Sci.* **1997**, *52*, 4105–4109.

Dietrich, E.; Mathieu, C.; Delmas, H.; Jenck, J. Raney-Nickel Catalyzed Hydrogenations: Gas–Liquid Mass Transfer in Gas-Induced Stirred Slurry Reactors. *Chem. Eng. Sci.* **1992**, *47*(13–14), 3597–3604.

Dry, M. E. The Fischer-Tropsch Process: 1950–2000. *Catal. Today* **2002**, *71*(3–4), 227–241.

Du, Y.; Chen, X.; Li, L.; Wang, P. Characteristics of Methane Desorption and Diffusion in Coal within a Negative Pressure Environment. *Fuel* **2018**, *217*, 111–121.

Dudukovi´c, M. P.; Larachi, F.; Mills, P. L. Multiphase Catalytic Reactors: A Perspective on Current Knowledge and Future Trends. *Catal. Rev.* **2002**, *44*, 123–246.

El Sayed, H. A.; El Naggar, A. M. A.; Heakal, B. H.; Ahmed, N. E.; Abdel-Rahman, A. A. H. Deep Catalytic Desulphurization of Heavy Gas Oil at Mild Operating Conditions Using Self-Functionalized Nanoparticles as a Novel Catalyst. *Fuel* **2017**, *209*, 127–131.

Eliason, S. A.; Bartholomew, C. H. Reaction and Deactivation Kinetics for Fischer-Tropsch Synthesis on Unpromoted and Potassium-Promoted Iron Catalysts. *Appl. Catal. A.* **1999**, *186*(1–2), 229–243.

Epstein, M.; Petrie, D. J.; Linehan, J. H.; Lambert ,G. A.; Cho, D. H. Incipient Stratification and Mixing in Aerated Liquid–Liquid or Liquid–Solid Mixtures. *Chem. Eng. Sci.* **1981**, *36*, 784–787.

Evans, G. M.; Jameson, G. J.; Atkinson, B. W. Prediction of the Bubble Size Generated by a Plunging Liquid Jet Bubble Column. *Chem. Eng. Sci.* **1992**, *47*, 3265–3272.

Fan, L.-S.; Yang, G. Q.; Lee, D. J.; Tsuchiya, K.; Luo, X. Some Aspects of High-Pressure Phenomena of Bubbles in Liquids and Liquid—Solid Suspensions. *Chem. Eng. Sci.* **1999**, *54*, 4681–4709.

Fan, L.-S.; Muroyama, K.; Chern, S. H. Hydrodynamics of Inverse Fluidization in Liquid-Solid and Gas–Liquid–Solid Systems. *Chem. Eng. J.* **1982a**, *24*, 143.

Fischer, F.; Tropsch, H. The Synthesis of Petroleum at Atmospheric Pressures from Gasification Products of Coal. *Brennstoff-Chemie* **1926**, *7*, 97–104.

Fukuma, M.; Muroyama, K.; Yasunishi, A. Properties of Bubble Swarm in a Slurry Bubble Column. *J. Chem. Eng. Jpn.* **1987**, *20*, 28–33.

Gandhi, B.; Prakash, A.; Bergougnou, M. A. Hydrodynamic Behavior of Slurry Bubble Column at High Solids Concentrations. *Powder Technol.* **1999**, *103*, 80–94.

Ganguly, S. K.; Das, G.; Kumar, S.; Sain, B.; Garg, M. O. Mechanistic Kinetics of Catalytic Oxidation of 1-Butanethiol in Light Oil Sweetening. *Catal. Today* **2012**, *198*(1), 246–251.

Gao, J.; Jia, C.; Li, J.; Zhang, M.; Gu, F.; Xu, G.; Zhong, Z.; Su, F. Ni/Al$_2$O$_3$ Catalysts for CO Methanation: Effect of Al$_2$O$_3$ Supports Calcined at Different Temperatures. *J. Energy Chem.* **2013**, *22*(6) 919–927.

Gao, Y.; Meng, F.; Cheng, Y.; Li, Z. Influence of Fuel Additives in the Urea-Nitrates Solution Combustion Synthesis of Ni-Al$_2$O$_3$ Catalyst for Slurry Phase CO Methanation. *Appl. Catal. A* **2017**, *534*, 12–21.

Gavroy, D.; Joly-Vuillemin, C.; Cordier, G.; Fouilloux, P.; Delmas, H. Continuous Hydrogenation of Adiponitrile on Raney Nickel in a Slurry Bubble Column. *Catal. Today* **1995**, *24*(1–2), 103–109.

Grasemann, M.; Renken, A.; Kashid, M.; Kiwi-Minsker, L. A Novel Compact Reactor for Three-Phase Hydrogenations. *Chem. Eng. Sci.* **2010**, *65*, 364–371.

Hashemi, S.; Macchi, A.; Servio, P. Gas–Liquid Mass Transfer in a Slurry Bubble Column Operated at Gas Hydrate Forming Conditions. *Chem. Eng. Sci.* **2009**, *64*, 3709–3716.

Hattori, K.; Tanaka, Y.; Suzuki, H.; Kubota, H. Kinetics of Liquid Phase Oxidation of Cumene in Bubble Column. *J. Chem. Eng. Jpn.* **1970**, *3*(1), 72–78.

Heck, R. M.; Smith, T. G. Acetylene Hydrogenation in a Bubble Column Slurry Reactor. *Ind. Eng. Chem. Process Des. Dev.* **1970**, *9*(4), 537.

Hichri, H.; Accary, A.; Andrieu, J. Kinetics and Slurry-Type Reactor Modelling During Catalytic Hydrogenation of O-Cresol On Ni/SiO$_2$. *Chem. Eng. Process.: Process Intensif.* *30*(3), **1991**, 133–140.

Iliuta, I.; Larachi, F. Wet Air Oxidation Solid Catalysis Analysis of Fixed and Sparged Three-Phase Reactors. *Chem. Eng. Process.: Process Intensif.* **2001**, *40*, 175–185.

Ishibashi, H.; Onozaki, M.; Kobayashi, M. Hayashi, J. I.; Itoh, H.; Chiba, T. Gas Holdup in Slurry Bubble Column Reactors of a 150 T/D Coal Liquefaction Pilot Plant Process. *Fuel* **2001**, *80*, 655–664.

Jana, S. K.; Bhaskarwar, A. N. Gas Absorption Accompanied by Chemical Reaction in a System of Three-Phase Slurry-Foam Reactors in Series. *Chem. Eng. Res. Des.* **2011**, *89*, 793–810.

Jin, H.; Liu, D.; Yang, S.; He, G.; Guo, Z.; Tong, Z. Experimental Study of Oxygen Mass Transfer Coefficient in Bubble Column with High Temperature and High Pressure. *Chem. Eng. Technol.* **2004**, *27*, 1267–1272.

Jin, H.; Yang, S.; He, G.; Liu, D.; Tong, Z.; Zhu, J. Gas–Liquid Mass Transfer Characteristics in a Gas–Liquid–Solid Bubble Column under Elevated Pressure and Temperature. *Chin. J. Chem. Eng.* **2014**, *22*, 955–961.

Kantarci, N.; Borak, F.; Ulgen, K. O. Bubble Column Reactors. *Process Biochem.* **2005**, *40*, 2263–2283.

Kluytmans, J. H. J.; Van Wachem, B. G. M.; Kuster, B. F. M.; Schouten, J. C. Mass Transfer in Sparged and Stirred Reactors: Influence of Carbon Particles and Electrolyte. *Chem. Eng. Sci.* **2003**, *58*, 4719–4728.

Köhler, E. O. Catalytic Dewaxing with Zeolites for Improved Profitability of ULSD Production. In *Studies in Surface Science and Catalysis*; Ruren, X.; et al., Eds.; Elsevier: The Netherlands, 2007; Vol. 170, pp 1292–1299.

Kolaczkowskia, S. T.; Plucinskia, P.; Beltranb, F. J.; Rivasa, F. J.; McLurgha, D. B. Wet Air Oxidation: A Review of Process Technologies and Aspects in Reactor Design. *Chem. Eng. J.* **1999**, *73*(2), 143–160.

Kölbel, H.; Klötzer, D.; Hammer, H. Zur Reaktionstechnik von Blasensäulen-Reaktoren mit suspendiertem Katalysator bei erhöhtem Druck. *Chem. Ing. Tech.* **1971**, *43*, 103–111.

Krämer, M.; Stöwe, K.; Duisberg, M.; Müller, F.; Reiser, M.; Sticher, S.; Maier, W. F. The Impact of Dopants on the Activity and Selectivity of a Ni-Based Methanation Catalyst. *Appl. Catal. A* **2009**, *369*(1–2), 42–52.

Kumar, S.; Kumar, R. A.; Munshi, P.; Khanna, A. Gas Hold-Up in Three Phase Co-Current Bubble Columns. *Proc. Eng.* **2012**, *42*, 782–794.

Ledakowicz, S.; Nettelhoff, H.; Kokuun, R.; Deckwer, W. D. Kinetics of the Fischer-Tropsch Synthesis in the Slurry Phase on a Potassium-Promoted Iron Catalyst. *Ind. Eng. Chem. Process. Des. Dev.* **1985**, *24*(4), 1043–1049.

Li, H.; Prakash, A.; Margaritis, A.; Bergougnou, M. A. Effects of Micron-Sized Particles on Hydrodynamics and Local Heat Transfer in a Slurry Bubble Column. *Powder Technol.* **2003**, *133*, 171–184.

Li, W.; Zhong, W. CFD Simulation of Hydrodynamics of Gas—Liquid—Solid Three-Phase Bubble Column. *Powder Technol.* **2015**, *286*, 766–788.

Li, Y.; Liu, M.; Li, X. Single Bubble Behavior in Gas—Liquid—Solid Mini-Fluidized Beds. *Chem. Eng. J.* **2016**, *286*, 497–507.

Li, X.; Chen, W.; Ma, L.; Wang, H.; Fan, J. Industrial Wastewater Advanced Treatment via Catalytic Ozonation with an Fe-Based Catalyst. *Chemosphere* **2018**, *195*, 336–343.

Liedtke A.-K.; Bornette, F.; Philippe, R.; de Bellefon, C. Gas–Liquid–Solid "Slurry Taylor" Flow: Experimental Evaluation Through the Catalytic Hydrogenation of 3-Methyl-1-Pentyn-3-Ol. *Chem. Eng. J.* **2013**, *227*, 174–181.

Lu, X. X.; Boyes, A. P.; Winterbottom, J. M. Operating and Hydrodynamic Characteristics of a Co-Current Downflow Bubble Column Reactor. *Chem. Eng. Sci.* **1994**, *49*, 5719–5733.

Luck, F. A Review of Industrial Catalytic Wet Air Oxidation Processes. *Catal. Today* **1996**, *27*(1–2), 195–202.

Luo, X.; Lee, D. J.; Lau, R.; Yang, G.; Fan, L.-S. Maximum Stable Bubble Size and Gas Holdup in High-Pressure Slurry Bubble Columns. *AIChE J.* **1999**, *45*, 665–680.

Majumder, S. K. *Hydrodynamics and Transport Processes of Inverse Bubbly Flow*, 1st ed.; Elsevier: Amsterdam, 2016; p 192.

Majumder, S. K.; Kundu, G.; Mukherjee, D. Mixing Mechanism in a Modified Co-Current Downflow Bubble Column. *Chem. Eng. J.* **2005**, *112*, 45–55.

Marcì, G.; García-López, E.; Vaiano, V.; Sarno, G.; Palmisano, L.; Keggin Heteropolyacids Supported on TiO_2 used in Gas–Solid (photo)Catalytic Propene Hydration and in Liquid–Solid Photocatalytic Glycerol Dehydration. *Catal. Today* **2017**, *281*(Part 1), 60–70.

Marwana, H.; Winterbottom, J. M. The Selective Hydrogenation of Butyne-1,4-Diol by Supported Palladiums: A Comparative Study on Slurry, Fixed Bed, and Monolith Downflow Bubble Column Reactors. *Catal. Today* **2004**, *97*, 325–330.

Mathew, S. P.; Rajasekharam, M. V.; Chaudhari, R. V. Hydrogenation of p-Isobutyl Acetophenone Using a Ru/Al_2O_3 Catalyst: Reaction Kinetics and Modelling of a Semi-Batch Slurry Reactor. *Catal. Today* **1999**, *49*(1–3), 49–56.

Mena, P.; Ferreira, A.; Teixeira, J. A.; Rocha, F. Effect of Some Solid Properties on Gas-Liquid Mass Transfer in a Bubble Column. *Chem. Eng. Process.* **2011**, *50*, 181–188.

Mukherjee, R. N.; Bhattacharya, P.; Taraphdar, D. K. Studies on the Dynamics of Three Phase Fluidization. In *Fluidization and its Application;* Angelino, H., et al.; Eds.; Cepadues-Editions: Toulouse, **1974,** pp 372–379.

Nedeltchev, S. Theoretical Prediction of Mass Transfer Coefficients in Both Gas–Liquid and Slurry Bubble Columns. *Chem. Eng. Sci.* **2017**, *157,* 169–181.

Nguyen, T. M.; Jung, J.; Lee, C. W.; Cho, J. Effect of Asphaltene Dispersion on Slurry-Phase Hydrocracking of Heavy Residual Hydrocarbons. *Fuel* **2018**, *214*, 174–186.

Ohkawa, A.; Kusabiraki, D.; Kawai, Y.; Sakai, N. Some Flow Characteristics of a Vertical Liquid Jet System Having Downcomers. *Chem. Eng. Sci.* **1986**, *41*, 2347–2361.

Onozaki, M.; Namiki, Y.; Sakai, N.; Kobayashi, M.; Nakayama, Y.; Yamada, T.; Morooka, S. Dynamic Simulation of Gas–Liquid Dispersion Behavior in Coal Liquefaction Reactors. *Chem. Eng. Sci.* **2000**, *55*, 5099–5113.

Öztürk, S. S.; Schumpe, A.; Deckwer, W.-D. Organic Liquids in a Bubble Column: Holdups and Mass Transfer Coefficients. *AlChE J.* **1987**, *33*(9), 1473–1480.

Pereira, O.; Martín-Alfonso, J. E.; Rodríguez, A.; Calleja, A.; López de Lacalle, L. N. Sustainability Analysis of Lubricant Oils for Minimum Quantity Lubrication Based on their Tribo-Rheological Performance. *J. Cleaner Prod.* **2017**, *164*, 1419–1429.

Pollington, S. D.; Enache, D. I.; Landon, P.; Meenakshisundaram, S. R.; Dimitratos, N.; Wagland, A.; Hutchings, G. J.; Stitt, E. H. Enhanced Selective Glycerol Oxidation in Multiphase Structured Reactors. *Catal. Today* **2009**, *145*(1–2), 169–175.

Reilly, I. G.; Scott, D. S.; Debruijn, T. J. W.; Macintyre, D. The Role of Gas Phase Momentum in Determining Gas Holdup and Hydrodynamic Flow Regimes in Bubble Column Operations. *Can. J. Chem. Eng.* **1994**, *72*, 3–12.

Saeidi, S.; Amin, N. A. S.; Rahimpour, M. R. Hydrogenation of CO_2 to Value-Added Products—A Review and Potential Future Developments. *J. CO_2 Util.* **2014**, *5*, 66–81.

Sangnimnuan, A.; Prasad, G. N.; Agnew, J. B. Gas Hold-Up and Backmixing in a Bubble-Column Reactor Undercoal-Hydroliquefaction Conditions. *Chem. Eng. Commun.* **1984**, *25*, 193–212.

Shah, Y. T.; Kelkar, B. G.; Godpole, S. P.; Deckwer, W. D. Design Parameters Estimations for Bubble Column Reactors. *AlChE J.* **1982**, *28*(3), 353–379.

Shah, Y. T.; Stiegel, G. J.; Sharma, M. M. Backmixing in Gas-Liquid–Solid Reactors. *AlChE J.* **1978**, *24*, 369–400.

Sivaiah, M.; Majumder, S. K. Gas Holdup and Frictional Pressure Drop of Gas–Liquid–Solid Flow in a Modified Slurry Bubble Column. *Int. J. Chem. React. Eng.* **2012,** *10*(Article A72), 1–29.

Sivaiah, M.; Majumder, S. K. Mass Transfer and Mixing in an Ejector-Induced Downflow Slurry Bubble Column. *Ind. Eng. Chem. Res.* **2013a,** *52*(35), 12661–12671.

Sivaiah, M.; Majumder, S. K. Hydrodynamics and Mixing Characteristics in an Ejector-Induced Downflow Slurry Bubble Column [EIDSBC]. *Chem. Eng. J.* **2013b,** *225*, 720–733.

Sivaiah, M.; Majumder, S. K. Dispersion Characteristics of Liquid in a Modified Gas–Liquid–Solid Three-Phase Down Flow Bubble Column. *Part. Sci. Technol.* **2013c,** *31*, 210–222.

Sivaiah, M.; Parmar, R.; Majumder, S. K. Gas Entrainment and Holdup Characteristics in a Modified Gas–Liquid–Solid Down Flow Three-Phase Contactor. *Powder Technol.* **2012,** *217*, 451–461.

Spiandore, M.; Souilah-Edib, M.; Piram, A.; Lacoste, A.; Doumenq, P. Desorption of Sulphur Mustard Simulants Methyl Salicylate and 2-Chloroethyl Ethyl Sulphide from Contaminated Scalp Hair After Vapour Exposure. *Chemosphere* **2018,** *191*, 721–728.

Strasser, W.; Wonders, A. Commercial Scale Slurry Bubble Column Reactor Optimization. *Adv. Fluid Mech. VII, WIT Trans. Eng. Sci.* **2008,** *59*, (WIT Press), 275–287.

Therning, P.; Rasmuson, A. Liquid Dispersion, Gas Holdup and Frictional Pressure Drop in a Packed Bubble Column at Elevated Pressures. *Chem. Eng. J.* **2001,** *81*, 331–335.

Usman, A. M.; Siddiqui, A. B.; Hussain, A.; Aitani, A.; Al-Khattaf, S. Catalytic Cracking of Crude Oil to Light Olefins and Naphtha: Experimental and Kinetic Modeling. *Chem. Eng. Res. Des.* **2017,** *120*, 121–137.

van Baten, J. M.; Ellenberger, J.; Krishna, R. Scale-Up Strategy for Bubble Column Slurry Reactors Using CFD Simulations. *Catal. Today* **2003,** 79–80, 259–265.

Vandu, C. O.; Koop, K.; Krishna, R. Volumetric Mass Transfer Coefficient in a Slurry Bubble Column Operating in the Heterogeneous Flow Regime. *Chem. Eng. Sci.* **2004,** *59*, 5417–5423.

Wang, H.; Zhang, J.; Bai, Y.; Wang, W.; Tan, Y.; Han, Y. NiO@SiO$_2$ Core-Shell Catalyst for Low-Temperature Methanation of Syngas in Slurry Reactor. *J. Fuel Chem. Technol.* **2016,** *44*(5), 548–556.

Xu, W.; Li, L. Y.; Grace, J. R.; Hébrard, G. Acid Rock Drainage Treatment by Clinoptilolite with Slurry Bubble Column: Sustainable Zinc Removal with Regeneration of Clinoptilolite. *Appl. Clay. Sci.* **2013,** 80–81, 31–37.

Xu, W.; Li, L. Y.; Grace, J. R. Zinc Removal from Acid Rock Drainage by Clinoptilolite in a Slurry Bubble Column. *Appl. Clay. Sci.* **2010,** *50*, 158–163.

Yang, W.; Wang, J.; Jin, Y. Mass Transfer Characteristics of Syngas Components in Slurry System at Industrial Conditions. *Chem. Eng. Technol.* **2001,** *24*, 651–657.

Yang, G.; Xu, Y.; Su, X.; Xie, Y.; Wang, J. MCM-41 Supported CuO/Bi$_2$O$_3$ Nanoparticles as Potential Catalyst for 1,4-Butynediol Synthesis. *Ceramics Int.* **2014,** *40*(3), 3969–3973.

Zhang, J.; Bai, Y.; Zhang, Q.; Wang, X.; Zhang, T.; Tan, Y.; Han, Y. Low-Temperature Methanation of Syngas in Slurry Phase Over Zr-Doped Ni/γ-Al$_2$O$_3$ Catalysts Prepared Using Different Methods. *Fuel* **2014,** *132*, 211–218.

CHAPTER 2

GAS DISTRIBUTION

CONTENTS

2.1 INTRODUCTION

The way of distribution of gas is a challenging engineering in industrial gas-aided slurry reactor design. The gas distributor is an important element in three-phase flow technology. It is designed to obtain the optimal yield during the operation. In practice, the distributors for gas distribution have taken a variety of forms based on the direction of gas entry either upward, laterally or downward. The suitability of the distributor depends on the process conditions, mechanical feasibility and cost.

The main requirements for a distributor are to:

- Boost homogeneous and stable flow distribution across the entire bed cross-section.
- Minimize the plugging.
- Minimize the attrition and weepage of bed particles.
- Minimize damage on erosion.
- Inhibit the flow-back of bed material during normal operation.

2.1.1 TYPE OF GAS DISTRIBUTORS

Different types of the gas distributors are suggested by the various authors to distribute the gas in the slurry reactor (Kunii and Levenspiel, 1991). Some important distributors are briefly discussed as follows:

2.1.1.1 LABORATORY SCALE DISTRIBUTOR

For small-scale (used for studies in the laboratory) the various types of distributor as shown in Figure 2.1 used generally are porous plate (ceramic and sintered), filter cloth, compressed fibres and compacted wire plate. Small-scale fluidization prefers these distributors because they have a sufficiently high flow resistance to give a uniform distribution of gas.

2.1.1.1.1 Porous Plate

The porous plates are generally made of stainless steel, copper, titanium, glass and alumina as shown in Figures 2.1a and 2.2a. The plates are prepared with pore sizes ranges 1–100 μm and with a porosity range of 0.25–0.75. For uniform gas flow, highly dispersion forms above the distributor, formation of little bubbles and emulsion, by coalescence bubbles rapidly move upward. Plates are furnished in a disc shape, rectangular shape or customer specified shape. Porous glasses with pore diameters ranging from some millimetres to around 20 μm can be obtained by sintering techniques. However, conventional sintering shows some disadvantages: the production of open pores with a diameter of only a few micrometre is difficult because such small pores can easily collapse caused by the viscous flow of the glasses during sintering. Before sintering, the glass powder is mixed with a salt of melting point (e.g. tricalcium phosphate, melting point at 1730°C) above the sintering temperature (between 1000 and 1520°C) of the glass and a high solubility in a solvent, in which the glass is insoluble. During the sintering process, the salt acts as a spacer. After cooling, it is leached and in dependence on the glass to salt ratio, porous glasses with an open porosity of up to 75% can be obtained. The diameters of the open pores depend on the grain size of the salt. Pyrogenic silicic acid (AEROSIL OX50) is used as a glassy phase in the salt sintering process.

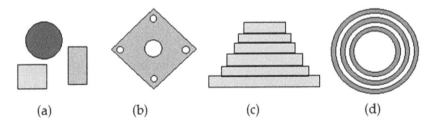

(a) (b) (c) (d)

FIGURE 2.1 Laboratory scale distributor: (a) Porous plate, (b) filter cloth, (c) compressed fibre and (d) compacted wire plate.

2.1.1.1.2 Filter Cloth

The filter cloths are made of cellulose, cotton-linter, cellulose-celite, polypropylene fabric, polyester fabric, cotton fabric and so forth. Different (round formats/disc) types of the cloths are commercially available in the market in various patterns and prints as shown in Figure 2.1b. They are manufactured as per the following types.

2.1.1.1.3   Multifilament or Continuous Filament

These are made by continuous extrusion of synthetic resins, which have a smooth surface are warped together to form the yarn using a S or Z turn. Multifilament yarns have a high-tensile strength.

2.1.1.1.4   Monofilament Type

These are made of single heavier extrusions. These extrusions are smooth, have high tensile strength and in some cases are modified with fillers and have excellent resistance to blinding.

2.1.1.1.5   Staple (Spun)-Type Distributor

These are prepared with chopped synthetic filaments by continuous extrusion. These short fibres are raked and twisted using a cotton or woollen system. These have the low tensile strength and a large surface area. Inhibition of the particle clogging is excellent.

2.1.1.1.6   Porous Stone Diffusers

These are used to disperse high volumes of fine bubbles for the application of wastewater treatment, to disinfection of potable water with ozone, stripping of volatile organic compounds in manufacturing processes and groundwater remediation sites and many others.

2.1.1.1.7   Ceramic Fine Bubble Diffuser

It comes in a variety of shapes and sizes ranging from domes and discs that operate in a vertical format to tubular or rod styles designed for use in a horizontal format. Materials of construction for ceramic diffusers range from the original porous slag to silica and the high-performance fused aluminium oxide materials. The most widely found use for ceramic diffusers is in biological treatment applications that require high mass transfer rates of oxygen interfacing with liquid such as wastewater treatment. For industrial operations, the laboratory scale distributors have some shortcomings as:

- High pressure drop leads to increased pumping power
- Low construction strength
- High cost for some materials
- Low resistivity against thermal stresses
- Clogging by file particles

2.1.1.1.8 Compressed Fibre

The diffuser is generally made of compressed glass, fibre, mineral or ozone resistant polymer fibre as shown in Figure 2.1c. These are designed by incorporating the matting or mesh impregnated with thermosetting resins that shrink on curing to produce the fine pores between spaces in the mesh or matting. It is generally preferable due to low cost, lower pressure loss, resistance to corrosive gas, resistance to biofilming, pore fouling, ease of bubble size and pressure loss alteration, thinness and lightweight (Action, 2013)

2.1.1.2 INDUSTRIAL SCALE DISTRIBUTOR

2.1.1.2.1 Perforated Plate Distributor

Perforated plate for distribution of process gas is a special one in a reactor. The perforations are formed as pores, which makes it possible to control the gas flow through the plate as shown in Figure 2.2.

In case of a large amount of solid with inlet gases, perforated plates without screen are preferred, for example, fluid catalytic cracking (FCC) reactor. Perforated plate cannot be used for high-temperature operation and highly reactive environment. The diameter of orifice in perforated plate ranges 1–2 mm in small-scale bed and to 50 mm in large-scale bed. The plates for the perforated distributors are standardized by different manufacturers as follows (Kunii and Levenspiel, 1991).

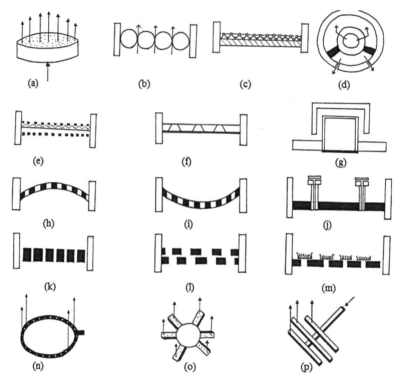

FIGURE 2.2   Industrial scale porous plate (not scaled): (a) perforated plate, (b) pipe grid, (c) ball type, (d) double pipe, (e) sandwich perforated, (f) multi-orifice, (g) bubble cap, (h) convex perforated, (i) concave perforated (j) multi-nozzle, (k) single perforated, (l) double perforated, (m) multi-filter, (n) ring type, (o) spider type and (p) multi-orifice bar.

Source: Redrawn by author freehand based Kunii and Levenspiel (1991).

2.1.1.2.1.1   Ring plates

For coil plates, a row of gills is punched in one operation.

2.1.1.2.1.2   Single-stroke plates

The holes on single-stroke plates are punched one at a time.

2.1.1.2.2   Ball-Type Distributors

These distributors are suitable for the bed where the dust-laden gases are used.

2.1.1.2.2.1   Gill plate

It is tailor-made for each processing plant to ensure gas distribution over the fluid bed, good transport of lumps and effective emptying of the fluid bed at shutdown.

2.1.1.2.2.2   Flex plate

It is designed in different patterns and shapes with sufficient strength to protect against vibrations during the operation.

2.1.1.2.2.3   Bubble cap plate

It is the latest process for gas dispersion. It does not allow the heat sensitive products to fall through the plate perforations. The distributor controls a solid movement in the reactor by the gas flow. Two types of bubble plates: one type is where the gills have one opening/hole as the traditional plates but with a special shape and another type is where the gills have an opening/hole in each end but in a different size.

2.1.1.2.2.4   Sandwich plate

Two perforated plates are sandwiched with a metal screen. These are stronger, easier to make, strip used to give a level surface for attaching wallboard. These prevent solid raining through the orifice.

2.1.1.2.2.5   Inclined plate

The inclined plate eliminates particle darning and vertical gas jet as shown in Figure 2.3a.

2.1.1.2.3   Nipple-Type Distributor

The nipple-type distributor is also suitably compared to multihole or orifice plate distributor because of it economic to fabricate and control of pressure drop by simply plugging the nipples at desired positions as shown in Figure 2.3d.

2.1.1.2.3.1   Screw-type distributors

These are simple to construct and can be a suitable compared to bubble cape distributor as shown in Figure 2.3e.

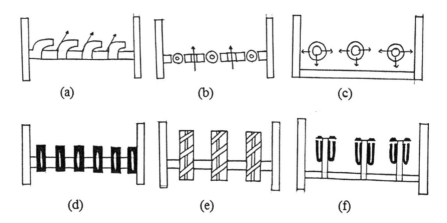

FIGURE 2.3   Industrial scale distributor (not scaled): (a) Inclined plate, (b) heat resistant grid type, (c) pipe grid type, (d) nipple plate and (e) screw plate.

Source: Redrawn by author freehand based on Kunii and Levenspiel (1991).

2.1.1.2.3.2   Staggered perforated plate (no screen)

In this type, there is no screen like a sandwich plate. It has lack of rigidity and deflects under heavy load. It needs reinforcing for support. There may be a gas leakage during the thermal expansion at the bed perimeter.

2.1.1.2.3.3   Dished perforated plate (no screen)

The Curved plate will withstand heavy loads and thermal stresses helps to counter the bubbling and channelling at the centre of the reactor, disadvantage for fabrication.

2.1.1.2.3.4   Grate bars (no screen)

Limited use, bars are two-dimensional version. The main disadvantages of the perforated plates are reactor leakage to plenum, high pressure drop and thermal distortion. It requires the peripheral seal to vessel shell, and support over long spans. Although the perforated plates have such disadvantages they are preferred because they can be easily fabricated, easy to modify hole size, easy to scale-up or down easy to clean, it can be flat, concave, convex or double dished, ports can be easily masked as shown in Figure 2.2.

2.1.1.2.4 Nozzle-Type Distributor

These are used for high-temperature operation and highly reactive environment. These are more expensive than the perforated plate. Multiple nozzle-type tuyere gives a good distribution above nozzle mouth as shown in Figure 2.2j. It is suitable for the equal gas flow fitted with high resistance orifice at its gas inlet. These are used for the preventing solids from falling through the distributor, to minimize the settlement of particles, sintering and stick on the distributor. At low gas flow rates bubble form and detach in orderly procession. At high velocities bubble coalescence into horizontal jets. Bubbles rise velocity exceeds linear growth rate. Low gas flow rates for $V_b = 1.138\, v^{6/5}/g^{3/5}$. Increasing orifice flow bigger bubbles form, the distance between successive bubbles decreases and coalescence to form a plumb. K value for orifice at height of 10 cm is 0.36 (bubbles rising with well-defined boundaries), at emulsion region it is 0.53 (voidage above emf) and at height of 25 cm it is 0.79. The nozzles are designed in a different way so that the jet created from the nozzle may be horizontally, upwardly or downwardly directed.

2.1.1.2.4.1 Bubble cap distributor

It has no orifice at its gas inlet. Orifices around the cape are designed to create sufficient pressure for uniform fluidization. It is not suitable for jetting effect of high-velocity gas which causes particle erosion by friction. Bubble form at bubble cap, at higher flow rates of gas, larger bubbles are generated without jet action as shown in Figure 2.2g.

2.1.1.2.4.2 Pipe distributor or sparger

For the larger bed, pipe grids distributors are generally recommended to use. The double pipe (Fig. 2.2d), pipe grid distributors as shown in Figure 2.3c have a special advantage over others such that it does not require any plenum chambers as shown in heat exchanger tubes as grid improves gas-liquid contact by breaking, growing bubbles, preventing gulf streaming, gross circulation of solids, reactant gas distributed by the grid along with carrier gas. At low gas flow rates a succession of bubbles and at high-velocities give standing flickering jet or plumb attached to the entry pipe. The pipe grid distributor can minimize weeping, well fit to multilevel fluid introduction. Down-pointing nozzles prevent clogging of the particle. Conical grids promote solid mixing, prevent stagnant solids build-up and minimize the solids segregations. Defluidizing solids under the grid have a major problem for this type of distributor. Different types of pipe distributor or sparger are shown in Figure 2.4. Further modification of the distributor can be based on the consideration of the extent of mixing of phases for the further dispersion in the reactor is shown in Figure 2.5.

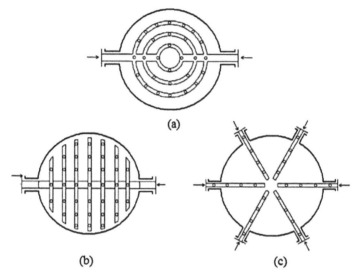

FIGURE 2.4   Different types of pipe distributor or sparger (not scaled): (a) multiple ring sparger, (b) spider and (c) pipe sparger.

Source: Redrawn based Kunii and Levenspiel (1991).

2.2   GAS DISTRIBUTION BY ENTRAINMENT IN A DOWNFLOW GAS-INTERACTING SLURRY COLUMN REACTOR

Entrainment of one phase into another by breaking a free liquid surface plays an important aspect in case of application of transport processes. It is found in practice in industrial as well as in environmental processes. There are various mechanisms to entrain gas in the processes. Some of the examples of gas entrainment mechanism to produce interfacial area to execute the transport process are entrainment of gas into water in open channels by wave circulation, atomization of gas by venturi, gas entrainment by thrusting into solid bodies or by plunging liquid jet through a free liquid surface. Gas entrainment by plunging liquid jets into a pool of liquid depends on the jet characteristics whether it is a free jet or wavy jet. The entrainment of gas significantly contributes to the total use of gas for different chemical and biochemical processes depending on the mechanism of entrainment.

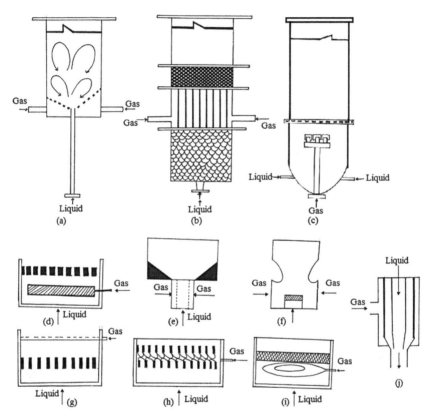

FIGURE 2.5 Other different types of distributor (not scaled): (a) spout mixing: (b) shell and tube type, (c) bubble cape in liquid plenum, (d) gas mixing in liquid plenum, (e) annular grid for gas, (f) venture mixer, (g) pipe grid gas inlet above perforated plate, (h) gas-liquid mixing in sandwich grid, (i) ring gas sparger below a wire mesh grid and (j) ejector-type sparger.

Source: Redrawn by author freehand based Kunii and Levenspiel (1991).

Gas entrainment by plunging liquid jets may be suitable to achieve enhancement of physical and reactive mass transfer united with intense mixing in slurry reactors. In particular, due to the favourable energy requirements, it can be used for the potential application in many chemical, fermentation and waste treatment processes. Gas entrainment by plunging liquid jets is also applied in flotation processes for mineral beneficiation. Gas entrainment by jets affords a good contact and dispersion of gas bubbles into the body of water which plays a significant role in this regard. The gas entrainment and its use for gas-interacting slurry reactor may be suitable for attaining better transport operations. Of course, mode of gas entrainment plays an important role to influence the gas hold-up, interfacial area and the degree of mass transfer in the gas-interacting slurry reactor.

Plunging jet system makes a high momentum to enhance the liquid mixing and gas distribution in the reactor. Since 1970, the interest and the scope of research on the plunging liquid jet systems are growing remarkably (McCarthy et al., 1970; Cumming, 1975; McKeogh et al., 1981; Bin, 1988, 1993; Evans et al., 1996; Sivaiah et al., 2012; Sivaiah and Majumder, 2013a–d; Majumder, 2016). Gas entrainment and its distribution by plunging liquid jets are controlled by jet velocity. The exclusive gas entrainment by liquid jet follows only when the relative velocity between the accelerate front and the liquid layer is greater than a certain minimum value. The gas entrainment occurs by two balancing mechanisms: (1) the interfacial shear along the liquid jet interface which carries down an air boundary layer and (2) the air entrapment process at the point of impact of the plunging jet with the receiving pool (Sivaiah et al., 2012). The entrainment is instigated by contacts between the interfaces and the turbulent eddies at the surface. These contacts ground surface distortions. The nature and degree of contacts are determined by the gravity, surface tension and the kinetic energy at the surface. At a very low-velocity and viscous (laminar) jets, an opposite of the dynamic bow developed between the jet and the unloading pool liquid which results in gas entrainment as shown in Figure 2.6. With further increasing the jet velocity an unbalanced gas film captured at the plunging point, whose length varies with the fluid properties and jet kinetic energy. The unbalanced gas film breaks up to generate bubbles, by contacts between the jet and the free surface. The bubble population generated depends on the physical properties of the liquid. In case of low viscous jets with conflicts on their surfaces, surfs of increasing amplitude move down the gas annulus surrounding the plunging point and intermittently tie it, which results in an entrainment in the form of a packed bubble group (Bin, 1993).

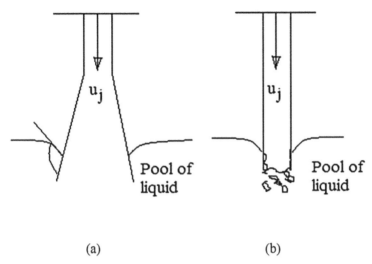

(a)   (b)

FIGURE 2.6   Mechanism of plunging of jet in open surface: (a) contact angle less than 180° and (b) air film formation.

Source: Redrawn by author based Majumder, 2016.

At viscosities less than 6.5 MPa, the gas annulus ceases to exist and consequence the trumpet-shaped zone around the jet (Cumming, 1975; Kusabiraki et al., 1990). The entrainment does not happen until the jet velocity increases further. For low viscous liquid, jets have disorders on their surfaces, and hence, the entrainment mechanism is directed by the interaction of these disorders with the surface of the unloading pool as shown in Figure 2.7. At first, the liquid surface forms a small dipped dynamic reversed meniscus which is produced by the impact pressure of the allied boundary layer gas and the flow field resulted itself in the pool liquid. When the surface is deformed by the sunken dynamic inverted meniscus, a transverse bulk liquid movement results in gas entrainment at the plunging zone. The successive disturbance may results in irregular entrainment. The surface irregularities become bigger with longer jets which results in higher entrainment. At minimum jet velocity that essential for entrainment, the jets yield a concave cut into a surface of the pool (Evans, 1996).

This is called an induction trumpet. The entrainment by the trumpet is more even than under the ultimate conditions. In case of ejector system, a free surface vortex occurs. In this case, significant oscillations of the concave cut geometry (induction trumpet) are overlaid on the smaller high-frequency oscillations of the throat length.

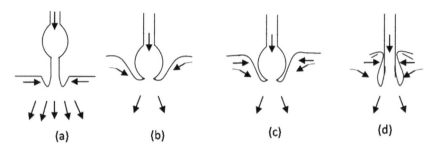

FIGURE 2.7   Gas entrainment by jets (a)—(d) subsequent phases of the occurrence when the jet moves downward.

Source: Redrawn by author based Majumder, 2016.

Cumming (1975) stated that 'tearing of the surface' arises from the liquid entrainment in the bulk at the tearing point is significantly assisted by the physical pressure downwards on the meniscus. The contribution of the gas boundary layer becomes more substantial. He also stated that small changes in the jet velocity can penetrate into the pool of liquid and the capture of gas is possible in almost stable condition. It depends on the angular momentum of the incoming liquid jet near the surface in the plunging zone. If it causes to intense, deep free vortices at the plunging zone around jets, enough entrainment of gas will not happen. The entrainment by plunging jet with downcomers with the delimited geometry and with the formation of the induction trumpet is not attractive. The trumpet steadily develops until a spurt of gas is carried into the bulk liquid (McCarthy et al., 1970).

High-velocity jet leads to the considerable quantities of gas entrainment. At this condition, the gas rises immediately around the plunging point by producing a setback of the flow in the unloading pool and outward circulation from the jet towards the walls without forming induction trumpet. This occurrence also results in the open liquid surface closely nearby to the jet being brushed perfect of low-density impurities and the jet retains its efficiency under polluting conditions. At high jet velocity gas entrainment occurs more regularly with formation of fine bubbles. At this condition, friction forces of gas develop noticeably and affect the roughness of the jet surface by encompassing the gas within the confines of the rough surface linked to the boundary layer. When the jet sways the pool surface both the encompassed gas and the boundary layer gas are carried underneath the open liquid surface. Thomas et al. (1984) suggested another mechanism at high jet velocities. They reported that maximum of the entrained gas arrives the main flow to the surface of the unloading fluid. In this regard, impact angle of the jet is the dominating factor for the gas entrainment. At sheer plunge, gas entrainment is less important due to the recirculating flows are relatively unruffled. Long jets with delay fail their consistency.

The gas entrainment depends on the basic hydrodynamic features of liquid jet such as jet diameter, jet velocity and physical properties of the fluids. Many other factors (nozzle design, jet inclination, vibration of jet, ejector configuration, jet penetration length and so forth.) can also affect jet characteristics.

2.3 CHARACTERISTICS OF LIQUID JETS

2.3.1 JET PARAMETERS

The jet diameter and the jet velocity are called as jet parameters, which should be estimated at the plunging point of the liquid jet. The gravity forces affect the diameter and velocity of the free falling jet. Free falling liquid jets are affected by contraction or expansion of the jet at the nozzle outlet from which jet is formed. From the energy balance, ignoring the effects of the adjoining gas and the minor contribution of the surface energy, the equation for jet diameter can be deduced as

$$\frac{d_j}{d_r} = \left(\frac{\pi^2 g L_j d_r^4}{8 Q_{sl}^2} + 1 \right)^{-1/4} \tag{2.1}$$

where d_r is the jet diameter at the reference point (Lin, 1969). For turbulent jets, jet velocity can be calculated as (Ciborowski and Biń, 1972)

$$u_j = \sqrt{u_o^2 + 2 g L_j} \ . \tag{2.2}$$

In this case, there may be a chance of developing a jet turbulence for which a loss of energy may vary from the actual values of jet velocity. The equation can be used for inclined jet, which forms a parabola, by replacing the height of the nozzle outlet above the pool liquid surface. For high-velocity jets, the jet surface becomes imprecise. The jet ranges in a distributed conical form. In this case, the jet diameter can be estimated by the equation represented by

$$\frac{d_j}{d_n} = C_{2.3}(We_g \, Re_{length})^{m_{2.3}} \qquad (2.3)$$

where $C_{2.3}$ is the numerical constant which is equal to 0.125 and $m_{2.3}=1/6$, valid for $We_A Re_{length}>7\times10^5$ and for jets created from long cylindrical nozzles ($l/d_n \geq 50$) (van de Sande and Smith, 1973) depending on the design of the nozzle or the established turbulence in the jet. The gas entrainment is affected by the nature of the jet coming from the different types of nozzles, which can be expressed in terms of dimensionless group $\mu_l / \sqrt{\rho_l \sigma L_j}$ (Ohkawa et al., 1987; Kusabiraki et al., 1992; Bin, 1993).

2.3.2 JET BREAK-UP LENGTH

Jet stability is one of the most important factors for gas entrainment by plunging liquid jets because gas entrainment is resulted by one mechanism of which individual droplets impinge on the liquid surface. The entrainment of gas depends on the break-up length of the jets whose stability differs at various regions at different flow conditions. Typically, the functionality of the break-up length, $L_B=f(u_o)$, has three regions such as (i) a laminar jet region, (ii) a turbulent region and (iii) a Weber number controlled region. For laminar jet region (Grant and Middleman, 1966)

$$L_B / d_n = 19.5We_o^{0.425}(1+3Z)^{0.85} \qquad (2.4)$$

where Z is called Ohnesorge number ($\mu_l / \sqrt{\rho_l \sigma L_j}$) (Ohnesorge, 1936) and the correlation is valid for $We_o^{0.5}(1+3Z)=3-100$. The break-up length for the turbulent region is

$$L_B / d_n = C_{2.5}We_o^{m_{2.5}} \qquad (2.5)$$

where the exponent $m_{2.5}$ ranges 0.31—0.32 for jets delivering from nozzles with $l/d_n>5$ (Bin, 1993). The transition from the laminar region to the turbulent is related to the Ohnesorge number and the nozzle geometry (Iciek, 1982). The gravity has influence on the jet diameter. Gravity accelerates the jet and, influences its break-up length. The break-up length of a contracting jet is longer than that of a cylindrical jet at the same initial conditions. According to Takahashi and Kitamura (1972), the break-up length for jets formed from nozzles with inner diameters ranging from 0.87 to 4.13 mm, and with L/d_n ratios from 9.3 to 57.0 for laminar low-viscosity jets is expressed as

$$L_B / d_n = 18(g\rho_l d_n^2 / \sigma)^{1/7} We_o^{3/7} \tag{2.6}$$

As per Van de Sande and Smith (1976), the break-up length is related to the Weber number as

$$L_B / d_n = 2.7 We_o^{0.5} \tag{2.7}$$

where We_o is expressed by the jet velocity. The break-up length of the jet is dependent on the instability strength in the jet. The break-up length with the release rate of liquid from the nozzle can be correlated as (McKeogh and Ervine, 1981)

$$L_B = C_{2.8} Q_g^{m_{2.8}} \tag{2.8}$$

The power exponent $m_{2.8}$ dependents on the instability strength in the jet without any change in the nozzle diameter.

2.3.3   VELOCITY PROFILE OF BOUNDARY LAYER OF A SURROUNDING GAS OF LIQUID JET

A boundary layer will develop along the jet when it is released from the nozzle. Different theoretical models appropriate to interpret the profile of the boundary layer of a surrounding gas or liquid jet are available in the literature (Sakiadis, 1961; Bin, 1974; Nguyen and Evans, 2006; Decent, 2008; Ma et al., 2010; Kendil et al., 2012). The velocity profile in the gaseous boundary layer should satisfy the following conditions (Bin, 1993): the momentum flux can be expressed as:

for $y = 0$:

$$\frac{\partial^2 u_g}{\partial y^2} = \frac{1}{r_n} \frac{\partial u_g}{\partial y} = 0 \tag{2.9}$$

for $y = \delta_i$:

$$\frac{\partial^2 u_g}{\partial y^2} = \frac{\partial u_g}{\partial y} = 0. \tag{2.10}$$

The integral balance of the momentum flux will yield the following equation:

$$\frac{d}{dx} \int_0^{\delta_i} u_g^2 2\pi r \, dr = v_g u_o / \beta \tag{2.11}$$

where the parameter β depends on the boundary layer thickness. The following velocity profile can be derived with the conditions of Equations 2.9 and 2.10:

$$\frac{u_g}{u_o} = 1 - \frac{2}{\beta} \ln\left(\frac{2r}{d_j}\right) + \frac{2}{\beta^3} \ln^3\left(\frac{2r}{d_j}\right) - \frac{1}{\beta^4} \ln^4\left(\frac{2r}{d_j}\right) \tag{2.12}$$

Substitution of this velocity profile (Eq. (2.12)) in Equation 2.11 yields the equation for parameter β, depending on a curvature parameter, ξ. According to Bin, 1974, the following correlations can be used

$$\beta = 0.027\xi^3 - 0.230\xi^2 + 1.158\xi \qquad for \quad 0 \le \xi \le 3 \tag{2.13}$$

$$\beta = 0.5517 + 3.3755 \log_{10} \xi \qquad for \quad \beta \ge 3 \tag{2.14}$$

The volumetric flow rate of dragged gas in the boundary layer by the liquid jet is then calculated as:

$$Q_g = A_d u_g \tag{2.15}$$

where A_d is called the displacement area which is defined as:

$$A_d = \frac{1}{u_o} \int_0^\infty (u_o - u_g) 2\pi r \, dr \tag{2.16}$$

$$A_d = (0.028\xi^3 + 0.195\xi^2 + 0.702\xi)\pi d_j^2 / 4 \qquad for \quad 0 \le \xi \le 1.1 \tag{2.17}$$

the correlation is applicable for the laminar boundary layers at $Re_{length} < 5 \times 10^5$.

2.4 GAS ENTRAINMENT RATE

A certain minimum value of jet velocity is required to plunge the jet through a liquid surface to entrain the gas. Estimation of the gas entrainment can be done by two methods (Sivaiah et al., 2012) which are (i) catching gas after it is entrained into the pool of liquid and (ii) measuring the removal of gas from a gaseous space above the pool surface in the separator. The first method belongs to the experiments with inclined jet and those in which bubbles traps are used. In the second method, the gas entrained in the space in the pool of the liquid in the vicinity of the plunging point is allowed to pass through an appropriate device, namely gas rotameter or by other instruments. The gas entrainment can be measured by vertical jets with the help of a collar mounted at the free liquid surface (van de Donk, 1981). High-speed photography technique is used to investigate the entrainment characteristics of such liquid jets varying their physical properties. Lin and Donnelly (1966) investigated the characteristics of gas entrainment by both laminar and turbulent jets. The mechanisms governing the entrainment process of the two types of jets are different those are described in the earlier section. The rate of gas entrainment in the gas-interacting ejector induced slurry column can be estimated a corresponding to the volumetric flow rate of slurry as:

$$\varepsilon_g = \frac{Q_g}{Q_g + Q_{sl}} \qquad (2.18)$$

By simplifying the Equation 2.18 one gets:

$$Q_g = \frac{Q_{sl}\varepsilon_g}{\left(1-\varepsilon_g\right)} \qquad (2.19)$$

where Q_g is the gas entrainment rate in the column. Q_{sl} is the volumetric liquid slurry flow rate and ε_g is the overall gas hold-up in the column which is defined as the fraction of volume of gas out of total gas-slurry mixture. Gas entrainment mainly depends on the volumetric flow rate of slurry or jet velocity and physical properties of slurry. Sivaiah et al. (2012) reported the profile of the gas entrainment ratio varying with the jet velocity in the gas-interacting downflow slurry reactor system. A correlation has been developed by Sivaiah et al. (2012) with the gas-liquid-solid phase system. From the experimental data, it is seen that the gas entrainment rate is a function of the physical, dynamic variables of the system which can be represented as:

$$\frac{Q_g}{u_j d_n^2} = 6.559 \times 10^{-18} \left(\frac{\sigma_l}{u_j^2 d_n w} \right)^{-0.085} \left(\frac{\mu_{eff}}{u_j d_n \rho_{sl}} \right)^{-4.655} \left(\frac{g d_n}{u_j^2} \right)^{1.349} \tag{2.20}$$

The range of gas entrainment ratio in downflow gas-interacting slurry reactor (DGISR) is reported by them is 0.03–0.91 at a range of Froude number 16–64. To explain the effects of nozzle geometry and physical properties of the liquid phase on the entrainment rate, three regions of the entrainment rate curve can be established (Ohkawa et al., 1987; Kusabiraki et al., 1990; Bin, 1993): (i) low jet velocity region; (ii) transition region and (iii) high jet velocity region. The usual S-shaped entrainment rate curve can be obtained for the entire range of the jet velocities. The low jet velocity region ranges up to about 5 m/s, whereas for the high jet velocity region at a value of $We_g > 10$ (van de Sande and Smith, 1973, 1976). The transition to these regions is subject to the experimental conditions and the physical properties of the system. To predict gas entrainment rates in the low and high-velocity regions different correlation models and mechanistic models can be established from experimental results. For the low-velocity region of entrainment, it can be assumed that individual bubbles form because of the surface tension forces averting the receiving flows from following the conflicts on the jet. In this regard, the gas entrainment rate is strongly dependent on the turbulence strength in the jet for this region and can be correlated as $Q_g \propto u_j^3$. For the high jet velocity region, it can be assumed that an air layer set into motion by shear stresses at the surface of the jet. It is broken up into a sequence of bubbles by unstable waves on its surface. The numerical derivation of the gas entrainment rate can be obtained based on balancing the pressure gradient in the air layer by the shear forces wielded on the air by the unloading flows which may be expressed as:

$$Q_g \propto \left(\frac{\mu_g}{\rho_{sl} g \sin \alpha} \right)^{0.5} u_j^{1.5} d_j \tag{2.21}$$

This model also advises that at high jet velocity the entrainment rate should not be dependent on the surface tension of the liquid, unlike at low speeds. The largest scales of turbulence are restricted by the jet diameter and hence, the rate of air entrainment may be reduced. The effect of surface tension restrains the gas entrainment rates with high-velocity jets owing to a decrease in the thickness of the gas layer. Bin (1993) stated that the rate of gas entrainment is related to the jet length near the break-up length (at least 90% of this length), whereas beyond the break-up point it is directly related to the kinetic energy of the jet. For shorter jets a characteristic factor suggested by van de Sande and Smith (1975, 1976) can be considered to express the gas entrainment rate as

$$X = d_o^2 u_j^3 L_j^{1/2} (\sin \alpha)^{-3/2} \qquad (2.22)$$

and the correlation for gas entrainment can be established by regression with the experimental results as

$$Q_g = C_{2.23} X^{m_{2.23}} \qquad (2.23)$$

Jets from shorter length nozzles results lower rate of entrainment at the same value of X. High-velocity jets carry the arrested gas within the envelope and entrain it in the boundary layer that grows outside the envelope into the pool. Thus, the total amount of gas that entrained can be conjured of two parts:

$$Q_g = Q_{g1} + Q_{g2} \qquad (2.24)$$

The first part of the right-hand side of the Equation 2.24 represents the gas arrested by the roughness of jet which can be expressed as

$$Q_{g1} = \frac{\pi}{4} \left(d_j^2 - d_o^2 \right) u_o \qquad (2.25)$$

whereas, the second part is that pulled as a laminar boundary layer mounting along the jet can be written as

$$Q_{g2} = \int_{d_j/2}^{\infty} 2\pi r u_g \, dr \qquad (2.26)$$

where u_g, is the local velocity of the gas in the boundary layer at a radial distance from the jet axis. In this regard, it is to be noted that gas might be arrested by the jet roughness and the chance of additional entrainment from the gas boundary layer (Cumming, 1975). It should be emphasized that the contribution (typically, it is between 20–70% of the total) of the boundary layer to the entrained gas depends upon an experimental conditions.

2.5   PERFORMANCE OF THE PLUNGING JETS FOR GAS ENTRAINMENT

The performance of the plunging liquid jet is assessed by the ratio of gas entrainment rate and liquid flow rate which is called entrainment ratio. The entrainment ratio depends on the jet characteristics such as jet velocity, jet length, jet diameter, jet inclination and physical properties of the system of jet used. A sharp increase of gas entrainment ratio is observed with the low jet velocity, whereas further increase on jet velocity, the ratio decreases marginally into a minimum and then increasing with a further increase in jet velocity (Bin, 1993) for air–water system. The same trend is followed in case of slurry system with particle concentration less than 2% by volume. Bin (1993) also reported that the influence of the jet angle on this ratio cannot be avoided. In order to, quantify the influence of different constraints on the entrainment ratio for long vertical cylindrical jet, a correlation can be developed in terms of the Froude number based on jet velocity and the ratio of jet length and nozzle diameter (L_j/d_o) which can be expressed as

$$Q_g / Q_l = \lambda Fr_j^a (L_j / d_o)^b. \tag{2.27}$$

The parameters λ, a and b depend on the experimental conditions. The values are 0.04, 0.28 and 0.4, respectively for air-water system within a range of $L_j/d_o \leq 100$ and $Fr_j^{0.28}(L_j / d_o)^{0.4} \geq 10$. However, the entrainment ratio for self-aerated flows can be interpreted by general equation, which for plunging jets yield $Q_g / Q_l \propto Fr_j$. Short cylindrical nozzles results the entrainment ratio lower at the same values of the key variables (such as u_j, L_j, d_o). The correlation can be developed with other shape of nozzle (like conical, nozzle with a flat entry). Van de Donk (1981) reported that uniform plunging of jet results growing values of the entrainment ratio. He stated that inoculation of a small amount of gas (1–3%) into liquid that form the jet has a significant effect on the entrainment ratio, except any major variation in the pressure drop across the nozzle. He also reported that a warped turbulence promoter astride in the conical nozzle can entrain more gas even at the rate of a higher pressure drop. The mechanically induced turbulence in the nozzle affects in change of the entrainment ratio. The dispersion of jet is also an important factor for entrainment efficiency. The more dispersed jets entrain more volume of gas and utilize their energy more prolixly in the biphasic region. Oppositely, due to lower pressure drop, jets produced from rather big conical nozzles may be favoured for industrial practices. Ahmed (1974) reported that annular nozzles of a similar range of outer diameters as the cylindrical nozzles may not entrain remarkable amount of gas within the same jet parameters range as compared with cylindrical nozzles. Other parameters, which also vary the entrainment rate, are the jet kinetic energy and physical parameters of phases. The rate of entrainment may be directly related to the jet kinetic energy. A higher entrainment rate can be attained at the same jet kinetic energy for jets produced from short nozzles of larger diameter. The variations of physical parameters of both phases on the entrainment rate are very attention-grabbing. The liquid phase viscosity increase affects the mechanism of entrainment. The entrainment rate can be correlated with the dimensionless number of physical properties as $\mu / \sqrt{\rho \sigma d_o}$. The dimensionless number is called Ohnesorge number. For the different liquids, the entrainment rate changes due to the change of shape of the liquid jet with physical properties of the liquid. With inclined jets, the gas entrainment rate decreases with increasing liquid viscosity and increases with an increase of dimensionless number in the low-velocity region. For the high-velocity region of the entrainment rate does not significantly change (Kumagai and Endoh, 1982). In the case of the opening region of entrainment, the functionality satisfy as $Q_g \propto v^{-0.95}$, whereas in the case of the low-velocity region it is $Q_g \propto v^{0.11}$. Kusabiraki et al. (1990) expressed the variation of gas entrainment ratio in terms of the physical parameters of the Ohnesorge number $(\mu / \sqrt{\rho \sigma d_o})$. The functionality is related on the jet velocity and the ratio of the nozzle length and the nozzle diameter and other experimental operating conditions. There is no effect of the gas phase physical properties on the entrainment rate (Burgess et al., 1972; Yagasaki and Kuzuoka, 1979). However, according to the theoretical considerations of Sene (1988), the gas entrainment rate should be proportional to $\sqrt{\mu_g}$. The penetration depth of jet from the nozzle influences the entrainment rate. The increase in penetration depth increases the entrainment rate. The increase of penetration depth can be controlled by introducing vertical downcomers. Proper design of the downcomer grades inducing a downward velocity field give rise to increased bubble penetration depth above the controlling values attainable in the ordinary slurry jet system. Such a downflow slurry down-comer with gas entrainment by a plunging jet adapts to a relatively new type of gas-interacting concurrent slurry reactor. A small-size downcomer is used around the plunging point which may have a constraint for the phase dispersion in the downcomer (Ohkawa et al., 1986). Initially, the penetration depth increases with increasing jet velocity, while length of the jet decreases because of a result of improved free liquid surface level in the downcomer until the gas as dispersed bubbles spread to the lower end of the downcomer. A uniform bubbling can be achieved through the down-comer at this jet velocity. Each flow condition links to a critical downward liquid velocity. Yamagiwa et al. (1990) established that the downflow system, having a lengthier phase contacting section in comparison to that in the normal liquid jet system, can be run without any apprehension about worse efficiency with regard to gas entrainment.

2.6   MODEL FOR ESTIMATION OF FILM-WISE ENTRAINMENT BY A LIQUID JET

The gas entrainment rate comprises of two components, namely the gas surrounded within the actual jet diameter at the plunging zone and the gas enclosed within the film between the jet and the induction trumpet surface (Evans, 1990) and (Evans and Jameson, 1991). To characterize the rupture dynamics, it is the important to know the mechanism for film rupture. The gas thickness of the gas layer reduces radially outwards, according to the thickness profiles at the bottom of the air film explained by Tran et al. (2013). The gas layer thickness close to the free surface is large, whereas in the radial position, the film thickness is smallest because of which it most likely to ruin. This radial position can be measured from the measured arc length to the bottom of the gas film. Tran et al. (2013) pointed out that if the film thickness is smaller than the critical value, in the order of 100 nm, there will be a leading of the van der Waals force between the two surfaces of the film and immediate rupture of the film (Couder et al., 2005; Dorbolo et al., 2005; Thoroddsen et al., 2012). The entrained gas diminishes after the gas film ruptures towards the bottom to generate gas bubbles and leaves behind those that are much smaller in size. This gas entrainment process is called as Mesler entrainment (Esmailizadeh and Mesler 1986; Pumphrey and Elmore 1990). Tran et al. (2013) concluded that if the jet velocity is small, the volume of the small bubbles left behind may be as good as to that of the principal generated bubbles. They reported such accomplishments by experiment at jet velocity <0.5 m/s for silicone oils. However, for higher jet velocity, the volume of the principal bubbles is foremost and can be used to estimate the total volume of the captured gas. The induction trumpet profile is shown in Figure 2.8.

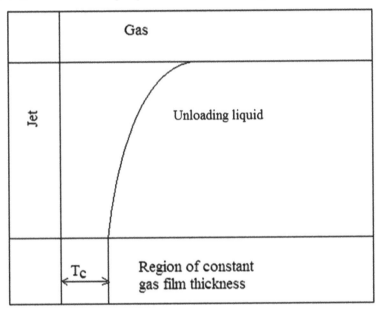

FIGURE 2.8   Profile of induction trumpet.

Source: Adapted from Sivaiah et al., 2012

Bin (1993) reported a model based on gas thin film theory for the estimation of the volumetric rate of film-wise entrainment by a liquid jet developed by Spiers et al., 1974. The model is based on the assumptions such as (i) a laminar velocity profile prevails in the film, (ii) the inside boundary velocity is equal to the jet velocity and (iii) the outside boundary velocity is equal to the recirculating eddy maximum velocity. The recirculating eddy is induced by the plunging liquid jet and its flow magnitude can be calculated from a correlation developed by Liu and Barkelew (1986) as

$$Q_{eddy,\max} = Q_l \left(\frac{0.37}{C_T} - 0.64 \right), \tag{2.28}$$

where C_T is the Craya–Curtet number (Barchilon and Curtet, 1964) where the secondary flow is zero which is defined as

$$C_T = \left[(d_c / d_j)^2 - 1/2\right]^{-1/2}. \tag{2.29}$$

The maximum recirculating eddy velocity is

$$u_{eddy,\max} = \frac{16 Q_{eddy,\max}}{\pi d_c^2}. \tag{2.30}$$

The volumetric flow rate inside the gas film can finally be calculated as follows:

$$Q_{g,in\ gas\ film} = \frac{\pi d_j^2}{2\mu_g} \left\{ \begin{array}{l} \frac{\rho_l g d_j^2}{64}(1-k^4) - \frac{\varphi}{2}[k^2 \ln(kd_j/2) - \ln(d_j/2)] \\ + \left(\frac{\varphi'}{2} - \frac{\varphi}{2}\right)(1-k^2) \end{array} \right\}, \tag{2.31}$$

where,

$$k = \left(r_j + T_c\right)/r_j \tag{2.32}$$

$$T_c = \sqrt{\frac{\mu_g(u_{eddy,\max})}{\rho_g g}}\psi(Ca_g). \tag{2.33}$$

The function ψ is a constant value of 0.21 for $Ca_g = (\mu_g u_{eddy,\max}/\sigma) \geq 3\times10^{-4}$.

$$\phi = \frac{4\mu_g(u_j - u_{eddy,\max}) + (1/4)\rho_l g d_j^2(1-k^2)}{4\ln k} \tag{2.34}$$

$$\phi' = -\left[\mu_g u_j + \frac{\rho_l g d_j^2}{16} + C_{48}\ln(d_j/2)\right] \tag{2.35}$$

The gas film thickness, T_c, can be evaluated from Equation 2.33 if the values of Q_g and d_j are known from experiments. Evans and Jameson (1991) attempted to develop a correlation for T_c. They reported that T_c is inversely proportional to d_c, which infer a negligible contribution of the film-wise entrainment by increasing the pool vessel size in diameter.

2.7   MINIMUM JET VELOCITY FOR GAS ENTRAINMENT

The entrainment of gas occurs at a certain minimum jet velocity at which slurry jet just plunges to the liquid pool in the column and pushes to break the surface and starts to encapsulate the gas as a bubble. At the minimum entrainment rate, the slurry jet contacts to the slurry pool surface and forms the first bubble in the column. The minimum jet velocity for the first gas bubble entrainment can be represented by a correlation as

$$u_{jm} = \left(\frac{\sigma_l}{C(\rho_{sl}\mu_{sl}d_n)^{0.5}} \right)^{2/3} \left[1 + C \left(\frac{\rho_g \mu_g L_j}{\rho_{sl}\mu_{sl}d_n} \right)^{0.5} \right]^{-2/3}, \qquad (2.36)$$

where C is a parameter which can be assessed from the experimental data. The value of C depends on the physical properties of the fluid, and jet characteristics. The value of parameter C in Equation 2.36 is found to be 0.234 for DGISR. From the experiment, it is very difficult to measure the length of the liquid jet. There is still no successful theoretical approach to predict the minimum entrainment velocity for the vertical plunging jets. From the experiment, the gas hold-up in the reactor can be easily estimated. Therefore, the estimation of the minimum entrainment velocity can be calculated from the profile of gas hold-up varying with the superficial slurry velocity. The superficial slurry velocity at which the gas hold-up value zero can be characterized as the minimum superficial slurry velocity (u_{slm}). The minimum jet velocity for the entrainment can be calculated as

$$u_{jm} = \frac{Q_{slm}}{(\pi/4)d_n^2} = u_{slm} \left(\frac{d_c}{d_n} \right)^2. \qquad (2.37)$$

The minimum jet velocity varies with the slurry concentration (Sivaiah et al., 2012) as

$$u_{jm} = \left(\frac{d_c}{d_n} \right)^2 e^{-\left[\frac{1.070-0.116w}{0.308-0.032w} \right]}, \qquad (2.38)$$

where, w is the slurry concentration (wt.%). The higher contracted slurry will result the higher value of minimum slurry velocity to entrain the minimum amount of gas in the reactor due to the higher resistance of penetration of jet into the pool of the slurry in the downcomer. The more concentrated slurry has higher viscosity. This specifies higher jet velocity required to get the same gas hold-up for the different slurry concentrations. For an air–water system without solid phase, the minimum jet velocity is found to be 3.099 m/s, whereas at a slurry concentration of 5%, it is 3.649 m/s. The onset value of the jet velocity at which gas entrainment starts is well-defined for coherent viscous laminar and turbulent jets (Ervine et al., 1980; McKeogh et al., 1981; Sene, 1988). Lin and Donnelly (1966) studied the gas entrainment characteristics and minimum jet velocity with a wide range of experimental conditions for laminar viscous jet and suggested an empirical correlation for $Re_j = 8$–1500, liquids of viscosities from 25 to 400 mPa·s, densities from 846 to 1246 kg/m³ and surface tensions from 0.03 to 0.063 N/m as

$$We_j = 10\,Re_j^{0.74}.$$

Detsch and Sharma (1990) reported that the minimum gas entrainment velocity for inclined jet can also be related to the ratio of dynamic pressure of the jet flow $(0.5\rho u_j^2)$ and the pressure force due to the surface tension $[\sigma(1/r_1 + 1/r_2)]$, where r_1, and r_2 are the radii of curvature of the water surface in the vertical and horizontal sections, respectively. One typical correlation developed by Detsch and Sharma (1990) in terms of the critical angle and the jet parameters is given as

$$\theta_c = \log_{10}[\sigma u_j / (\mu_l \rho_l)]^{-242.13} - 79.1 \tag{2.39}$$

the range of validity of Equation 2.39 is $0.01 < \sigma u_j / (\mu_l \rho_l) < 0.5$ m⁵/(kg·s²).

2.8   ENERGY EFFICIENCY OF GAS ENTRAINMENT

The gas entrainment is one of the key parameters of the gas-interacting downflow slurry reactor. The energy efficiency of the slurry reactor for gas entrainment can be assessed by the rate of kinetic energy supplied and the amount utilized by the jet plunging for the gas entrainment. The rate of energy supplied can be calculated by the equation (Bin and Smith, 1982) given by

$$\dot{E}_s = \frac{\pi \rho_{sl} d_n^2 u_j^3}{8}. \tag{2.40}$$

The utilize energy for the mixing and entrain the gas in the gas–liquid–solid mixture by penetrating the jet into the pool of the slurry can be calculated by the product of energy supplied to the system and the energy utilization efficiency of mixing, which can be expressed as

$$\dot{E}_{um} = \eta_m \dot{E}_s. \qquad (2.41)$$

At the entrainment region of the slurry, a complete capture of the entrained gas continues unless certain energy is executed on the gas bubbles to move downward (Bin, 1988). Thus, a minimum kinetic energy is essential to passage the gas bubbles to downward through the contactor. This minimum kinetic energy is an essential subject to the diameter of the column and the physical properties of the fluids. Based on the fluid properties a correlation suggested by Sivaiah et al. (2012) for the minimum energy required to entrain the gas in the pool of the slurry inside the contactor for the DGISR which can be expressed by

$$\dot{E}_{min} = 4.62 \times 10^3 \frac{\mu_{sl}^2 d_n}{\rho_{sl}} \left(\frac{d_n \sigma_l \rho_g}{\mu_g \mu_{sl}} \right)^{1.61}. \qquad (2.42)$$

The minimum energy requirement for the gas entrainment is 2.926×10^{-4} KW for air–water system. The rate of minimum energy required is higher for higher concentration of slurry. The higher resistance slurry is due to its apparent viscosity. The energy efficiency of the DGISR system and other types of gas–liquid contactors has been summarized in Table 2.1.

TABLE 2.1 Gas Entrainment Performance Based on Energy Supply.

Authors	Type of reactor	$\dot{E}_s / Q_{g\,[KW\ s/m3]}$
Fukuda et al. (1968)	Aerated—stirred fermenter	80–140
Topiwala and Hamer (1974)	Hollow impeller-based reactor	300–700
Zundelevich (1979)	Turbo aerator	60–800
Matsumura et al. (1982)	Tank-type gas-interacting reactor	100–1000
Ohkawa et al. (1986)	Water jet aeration in pool system	15–30
Sivaiah et al. (2012)	Downflow gas-interacting slurry reactor	101–210

KEYWORDS

- **gas distributors**
- **gas entrainment**
- **induction trumpet**
- **industrial scale distributor**
- **laboratory scale distributor**
- **liquid jets**
- **minimum jet velocity**

REFERENCES

Action, A. *Advances in Oxygen Research and Application;* Scholarly Edition™: Atlanta, Georgia, 2013.

Ahmed, A. Aeration by Plunging Liquid Jet. Ph.D. Thesis, Loughborough University of Technology, 1974.

Barchilon, M.; Curtet, R. Some Details of the Structure of an Axisymmetric Confined Jet with Backflow. *J. Basic Eng. Trans. ASME* **1964,** *86,* 777–787.

Bin, A. Verification of Different Theories of Gas Entrainment by Plunging Liquid Jets. *Inż. Chem. (Polish)* **1974,** *4,* 523–540.

Bin, A. K. Minimum Air Entrainment Velocity of Vertical Plunging Liquid Jets. *Chem. Eng. Sci.* **1988,** *43,* 379–389.

Bin, A. K. Gas Entrainment by Plunging Liquid Jets. *Chem. Eng. Sci.* **1993,** *48,* 3585–3630.

Bin, A. K.; Smith, J. M. Mass Transfer in a Plunging Liquid Jet Absorber. *Chem. Eng. Commun.* **1982,** *15,* 367–383.

Burgess, J. M.; Molloy, N. A.; McCarthy, M. J. A Note on the Plunging Jet Reactor. *Chem. Eng. Sci.* **1972,** *27,* 442–445.

Ciborowski, J.; Biń, A. Minimum Entrainment Velocity for Free Liquid Jets. *Inż. Chem. (Polish)* **1972,** *2,* 453–469.

Couder, Y.; Fort, E.; Gautier, C. H.; Boudaoud, A. From Bouncing to Foating: Noncoalescence of Drops on a Fluid Bath. *Phys. Rev. Lett.* **2005,** *94,* 177801.

Cumming, I. W. The Impact of Falling Liquids with Liquid Surfaces. Ph.D. Thesis, Lough. Univ. of Technology, 1975.

Decent, S. P. A Simplified Model of the Onset of Air Entrainment in Curtain Coating at Small Capillary Number. *Chem. Eng. Res. Des.* **2008,** *86*(3), (March) 311–323.

Detsch, R. M.; Sharma, R. N. The Critical Angle for Gas Bubble Entrainment by Plunging Liquid Jets. *Chem. Eng. J.* **1990,** *44,* 157–166.

Dorbolo, S.; Reyssat, E.; Vandewalle, N.; Quéré, D. Aging of an Antibubble. *Eur. Phys. Lett.* **2005,** *69,* 966–970.

Ervine, D. A.; McKeogh, E. J.; Elsawy, E. M. Effect of Turbulence Intensity on the Rate of Air Entrainment by Plunging Water Jets. *Proc. Inst. Civ. Eng. Part 2* **1980,** *69,* 425–445.

Esmailizadeh, L.; Mesler, R. Bubble Entrainment with Drops. *J Colloid Interface Sci.* **1986,** *110*(2), 561–574.

Evans, G. M. A Study of a Plunging Jet Bubble Column. Ph.D. Thesis, University of Newcastle: Australia, 1990.

Evans, G. M.; Jameson, G. J. Prediction of the Gas Film Entrainment Rate for a Plunging Liquid Jet Reactor. *Proceedings of the A.I. Ch.E. Symposium on Multiphase Reactors,* Flouston, Texas, 1991.

Evans, G. M.; Jameson, G. J.; Reilly, C. D. Free Jet Expansion and Gas Entrainment Characteristics of a Plunging Liquid Jet. *Exp. Therm. Fluid Sci.* **1996,** *12,* 142–149.

Fan, L. S. *Gas–Liquid–Solid Fluidization Engineering. Butterworth's Series in Chemical Engineering;* Butterworth-Heinemann: Stonehame, USA, 2004, 158–159.

Fukuda, H.; Sumino, Y.; Kanzaki, T. Scale-up of Fermenters. II. Modified Equations for Power Requirement. *J. Ferment. Technol.* **1968,** *46,* 838–845.

Grant, R. P.; Middleman, S. Newtonian Jet Stability. *AIChE J.* **1966,** *12,* 669–678.

Iciek, J. The Hydrodynamics of Free, Liquid Jet and Their Influence on Direct Contact Heat Transfer-I. Hydrodynamics of a Free, Cylindrical Liquid Jet. *Int. J. Multiphase Flow* **1982,** *8,* 239–249.

Kendil, F. Z.; Danciu, D. V.; Schmidt, M.; Salah, A. B.; Lucas, D.; Krepper, E.; Mataoui, A. Flow Field Assessment Under a Plunging Liquid Jet. *Prog. Nucl. Energy* (April) **2012,** *56,* 100–110.

Kumagai, M.; Endoh, K. Effects of Kinematic Viscosity and Surface Tension on Gas Entrainment Rate of an Impinging Liquid Jet. *J. Chem. Eng. Jpn.* **1982,** *15,* 427–433.

Kunii, D.; Levenspiel, O. *Fluidization Engineering,* 2nd ed.; Butterworth-Heinemann: E.U.A., **1991**; Vol. *82,* pp 105–106.

Kusabiraki, D.; Niki, H.; Yamagiwa, K.; Ohkawa, A. Gas Entrainment Rate and Flow Pattern of Vertical Plunging Liquid Jets. *Can. J. Chem. Eng.* **1990,** *68,* 893–903.

Kusabiraki, D.; Yamagiwa, K.; Yasuda, M.; Ohkawa, A. Gas Entrainment Behavior of Vertical Plunging Liquid Jets in Terms of Changes in Jet Surface Length. *Can. J. Chem. Eng.* **1992,** *70,* 181–184.

Lin, T. J.; Donnelly, H. G. Gas Bubble Entrainment by Plunging Laminar Liquid Jets. *AIChE J.* **1966,** *12,* 563–571.

Liu, C.; Barkelew, C. W. Numerical Analysis of Jet-Stirred Reactors With Turbulent Flows and Homogeneous Reactions. *AIChE J.* **1986,** *32,* 1813–1820.

Ma, J.; Oberai, A. A.; Drew, D. A.; Lahey, Jr., R. T.; Moraga, F.J. A Quantitative Sub-Grid Air Entrainment Model for Bubbly Flows—Plunging Jets. *Comput Fluids* **2010,** *39*(1), (January) 77–86.

Majumder, S. K. *Hydrodynamics and Transport Processes of Inverse Bubbly Flow,* 1st ed.; Elsevier: Amsterdam, 2016; p 192.

Matsumura, M.; Sakuma, H.; Yamagata, T.; Kobayashi, J. Gas Entrainment in a New Gas Entraining Fermenter. *J. Ferment. Technol.* **1982,** *60,* 457–467.

McCarthy, M. J.; Henderson, J. B.; Molloy, J. B. Gas Entrainment by Plunging Jets. *Proceedings of Chemeca Melbourne,* Section 2 1970, 86–100.

McKeogh, E. J.; Ervine, D. A.; Endoh, K. Air Entrainment Rate and Diffusion Pattern of Plunging Liquid Jets. *Chem. Eng. Sci.* **1981,** *36,* 1161–1172.

Nguyen, A. V. Evans, G. M. Computational Fluid Dynamics Modelling of Gas Jets Impinging Onto Liquid Pools. *Appl. Math. Modell.* **2006,** *30*(11), (November) 1472–1484.

Ohkawa, A.; Kusabiraki, D.; Sakai, N. Effect of Nozzle Length on Gas Entrainment Characteristics of Vertical Liquid Jet. *J. Chem. Eng. Jpn.* **1987,** *20*, 295–300.

Ohkawa, A.; Kusabiraki, D. Y.; Kawai, N. Sakai. Some Flow Characteristics of a Vertical Liquid Jet System Having Down Comers. *Chem. Eng. Sci.* **1986,** *41*, 2347–2361.

Ohnesorge, W. Formation of Drops by Nozzles and the Breakup of Liquid Jets. *J. Appl. Math. Mech.* **1936,** *16*, 355–358.

Pumphrey, H. C.; Elmore, P. A. Entrainment of Bubbles by Drop Impacts. *J. Fluid Mech.* **1990,** *220*, 539–567.

Sakiadis, B. C. Boundary-Layer Behaviour on Continuous Solid Surfaces-Ill. The Boundary Layer on a Continuous Cylindrical Surface. *AIChE. J.* **1961,** *8*, 467–472.

Sene, K. J. Air Entrainment by Plunging Jets. *Chem. Eng. Sci.* **1988,** *43*, 2615–2623.

Sivaiah, M.; Majumder, S. K. Mass Transfer and Mixing in an Ejector-Induced Downflow Slurry Bubble Column. *Ind. Eng. Chem. Res.* **2013a,** *52*(35), 12661–12671.

Sivaiah, M.; Majumder, S. K. Fluid Dynamics in Ejector-Induced Efficient Slurry Bubble Column. *Int. J. Innovation* **2013b,** *4*(3), 1–7.

Sivaiah, M.; Majumder, S. K. Hydrodynamics and Mixing Characteristics in an Ejector-Induced Downflow Slurry Bubble Column [EIDSBC]. *Chem. Eng. J.* **2013c,** *225*, 720–733.

Sivaiah, M.; Majumder, S. K. Dispersion Characteristics of Liquid in a Modified Gas-Liquid-Solid Three-Phase Down Flow Bubble Column. *Part. Sci. Technol.* **2013d,** *31*, 210–220.

Sivaiah, M.; Parmar, R.; Majumder, S. K. Gas Entrainment and Holdup Characteristics in a Modified Gas-Liquid-Solid Down Flow Three-Phase Contactor. *Powder. Technol.* **2012,** *217*, 451–461.

Spiers, R. P.; Subbaraman, C. V.; Wilkinson, W. L. Free Coating of a Newtonian Liquid Onto a Vertical Surface. *Chem. Eng. Sci.* **1974,** *29*, 389–396.

Takahashi, T.; Kitamura, Y. Stability of a Contracting Liquid Jet. Memoirs of the School of Eng. (Okayama University) 1972, 7, 61–84.

Thomas, N. H.; Auton, T. R.; Sene, K.; Hunt, J. C. R. Entrapment and Transport of Bubbles by Plunging Water. In *Gas Transport at Water Surfaces;* Brutsaert, W., Jirka, G. H., Eds.; Reidel: Dordrecht, 1984; pp 255–268.

Thoroddsen, S. T.; Thoraval, M. J.; Takehara, K.; Etoh, T. G. Micro-Bubble Morphologies Following Drop Impacts Onto a Pool Surface. *J. Fluid Mech.* **2012,** *708*, 469–479.

Topiwala, H. H.; Hamer, G. Mass Transfer and Dispersion Properties in a Fermenter with a Gas-Inducing Impeller. *Trans. Inst. Chem. Eng.* **1974,** *52*, 113–120.

Tran, T.; De Maleprade, H.; Sun, C.; Lohse, D. Air Entrainment during Impact of Droplets on Liquid Surfaces. *J. Fluid Mech.* **2013,** *726*, R3.

Van de Donk, J. A. C. Water Aeration with Plunging Jets. Ph.D. Thesis, Technische Hogeschool Delft: Netherlands, 1981.

Van de Sande, E.; Smith, J. M. Surface Entrainment of Air by High Velocity Water Jets. *Chem. Eng. Sci.* **1973,** *28*, 1161–1168.

Van de Sande, E.; Smith, J. M. Mass Transfer from Plunging Water Jets. *Chem. Eng. J.* **1975,** *10*, 225–233.

Van de Sande, E.; Smith, J. M. Jet Break-up and Air Entrainment by Low Velocity Turbulent Water Jets. *Chem. Eng. Sci.* **1976,** *31*, 219–224.

Yamagiwa, K.; Kusabiraki, D.; Ohkawa, A. Gas Holdup and Gas Entrainment Rate in Downflow Bubble Column with Gas Entrainment by a Liquid Jet Operating at High Liquid Throughput. *J. Chem. Eng. Jpn.* **1990,** *23,* 343–348.

Yagasaki, T.; Kuzuoka, T. Surface Entrainment of Gas by Plunging Liquid Jet. Research Reports of Kogakuin University 1979, 47, 77–85.

Zundelevich, Y. Power Consumption and Gas Capacity of Self-Inducting Turbo Aerators. *AIChE J.* **1979,** *25,* 763–773.

CHAPTER 3

GAS HOLD-UP CHARACTERISTICS

CONTENTS

3.1   DEFINITION OF GAS HOLD-UP

Gas hold-up is one of the key parameters describing the hydrodynamics of three-phase reactors. It is defined as the volume fraction of the gas in total volume of three-phase mixture in the gas-interacting slurry reactor which can be expressed as

$$\varepsilon_g = \frac{V_g}{V_t} = \frac{\text{Volume of gas}}{\text{Volume of gas–liquid–solid mixture}} \tag{3.1}$$

The gas hold-up is also expressed by the slip ratio (S) of the phase flow which can be written as

$$S = \frac{u_g}{u_{l-s}} = \frac{Q_g/(A\varepsilon_g)}{Q_{l-s}/(A(1-\varepsilon_g))} = \frac{Q_g(1-\varepsilon_g)}{Q_{l-s}\varepsilon_g} \tag{3.2}$$

$$S = \frac{u_g}{u_{l-s}} = \frac{Gx/(A\varepsilon_g\rho_g)}{G(1-x)/(A(1-\varepsilon_g)\rho_l)} = \frac{\rho_{l-s}x(1-\varepsilon_g)}{\rho_g(1-x)\varepsilon_g} \tag{3.3}$$

Equations 3.2 and 3.3 can be expressed in different forms:

$$\varepsilon_g = \frac{\dot{Q}_g}{S\dot{Q}_{l-s} + \dot{Q}_g} \tag{3.4}$$

$$\varepsilon_g = \frac{1}{1 + S\left(\dfrac{1-x}{x}\right)\left(\dfrac{\rho_g}{\rho_{l-s}}\right)} \tag{3.5}$$

The volumetric quality ($\varepsilon_g = \dot{Q}_g/(\dot{Q}_{l-s} + \dot{Q}_g)$) corresponds to the gas hold-up when the slip ratio (S) is unity. As the slip ratio is approached to 1, the homogeneity of the mixture is approached to perfect. If the ratio of liquid to gas density is large, the gas hold-up based on the homogeneous mixture of the phases increases very sharply once the mass quality increases even slightly above zero. The mass quality (x) is defined as the ratio of the gas mass flow rate to the total mass flow rate.

The distribution of gas hold-up results in fluid backmixing in the reactor. This backmixing has effect on the heat transfer and mass transfer. The gas hold-up profile in the reactors determines the flow regimes, degree of liquid mixing and the degrees of transport processes which are the important factor for scaling up the reactor. The area-average gas hold-up distributions can be represented by Equation 3.6.

$$\varepsilon_g(r) = \frac{1}{2\pi} \int_0^{2\pi} \varepsilon_g(r,\theta)d\theta \qquad (3.6)$$

where $\varepsilon_g(r)$ is the radial gas hold-up at a radial distance r. The hold-up profiles in the well-developed region do not depend on the axial position. There are a number of similar forms of empirical equations that can be fitted to the observed hold-up profiles. Nassos and Bankoff (1967) suggested a profile of a radial hold-up profile which can be expressed as

$$\varepsilon_g = \tilde{\varepsilon}_g \left(\frac{n+2}{n}\right)\left\{1-\left(\frac{r}{R}\right)^n\right\} \qquad (3.7)$$

The exponent n is a parameter which indicates the steepness of the hold-up profile and r/R is the dimensionless radial position. When n is large, the profile is flat and for small n, the profile is steep. If n is approximately equal to 2, the profile is parabolic. The steepness of the hold-up profile results in the intensity of liquid circulation. Ueyama and Miyauchi (1979) modified the above equation as follows to include the possibility of finite gas hold-up close to the wall.

$$\varepsilon_g = \tilde{\varepsilon}_g \left(\frac{n+2}{n}\right)\left\{1-c\left(\frac{r}{R}\right)^n\right\} \qquad (3.8)$$

where c is an additional parameter which is indicative of the value of gas hold-up near the wall. If $c=1$, there is zero hold-up close to wall, and if $c=0$, hold-up is constant with changing r/R. The hold-up $\tilde{\varepsilon}_g$ is the radial chordal average gas hold-up along the diameter of the reactor which is related to the cross-sectional average gas hold-up. The cross-sectional average gas hold-up is expressed by

$$\tilde{\varepsilon}_g / \overline{\varepsilon}_g = \frac{n}{n+2-2c} \qquad (3.9)$$

The cross-sectional average gas hold-up is also calculated by averaging the radial gas hold-up profiles as:

$$\overline{\varepsilon}_g = \frac{1}{\pi R^2} \int_0^R 2\pi r \varepsilon_g(r)dr \qquad (3.10)$$

The radial gas hold-up profile at atmospheric pressure is parabolic in nature at a gas velocity of 0.12 m/s, indicating churn turbulent flow, whereas at higher pressure, the profile is flatter (Kemoun et al., 2001).

3.2   IMPORTANCE OF GAS HOLD-UP

- The gas hold-up is an important designing parameter to predict the hydrodynamic behaviour of slurry bubble column reactors.
- The gas hold-up in the reactor determines the residence time of the slurry phase and the degree of slurry mixing.
- A low gas velocity results in a homogeneous bubble flow while a high gas velocity, a heterogeneous bubble flow or a churn-turbulent flow with rising bubbles; having a wide size distribution or irregularly distorted bubbles due to frequent bubble coalescence and/or break-up, this phenomena depends on the gas hold-up.
- The gas hold-up determines the controlling of the feeding device, properties of gas and liquid, pressure, temperature, column size, as well as the superficial gas velocity.
- The gas hold-up in the reactor further affects the slurry properties that are a function of liquid and solid properties and solid concentration.
- The flow pattern transition in a gas-aided slurry reactor shifts by the controlling of the gas hold-up in the reactor. The homogeneous regime is observed in column reactors at relatively low superficial gas velocities ($u_g < 0.05$ m/s) which are characterized by a uniform distribution of the gas hold-up.
- The radial gas hold-up distribution generates the turbulent liquid circulation, in turn, controls the liquid phase mixing and transport coefficient in deciding the performance of the reactor. Owing to this local maximum in the gas hold-up, the density of gas–liquid dispersion is less at the centre and more near the walls. This density difference is the driving force for the liquid circulation.
- The size distribution of the bubble in the gas-aided slurry reactor depends on the gas hold-up and particle concentration. The bubble particle and the bubble interaction are also governed by the gas hold-up in the slurry reactor.
- The overall mass and heat transport rate per unit volume of the dispersion in a gas-aided slurry reactor is governed by the liquid-side mass transfer and heat transfer coefficients. In a slurry bubble

column reactor, the variation in mass transfer coefficient is primarily due to variations in the interracial area. The specific gas–liquid interfacial area, depending on the gas hold-up and the mean bubble diameter, can be expressed as $a = 6\varepsilon_g/d_b$. Thus, a precise knowledge of the gas hold-up is needed to determine the specific gas–liquid interfacial area.

3.3   METHOD TO ESTIMATE THE GAS HOLD-UP

The design of gas–liquid–solid reactors still depends, to a large extent, on empirical modelling, which in turn are based on measurements data of laboratory or pilot plant basis. These correlation models are used to design and scale up the reactor. However, modern approaches such as computational fluid dynamics are used to help in the design of such reactors precisely. But validity and the reliability of the code depend on the experimental verification which is still lacking in their suitability. In this regard, the accurate measurement is quite challenging by using different techniques. Gas hold-up is measured by numerous invasive or non-invasive techniques, which have been reviewed by Kumar et al. (1997) and Boyer et al. (2002). Among these techniques, average gas hold-up measurement through pressure difference measurements and phase isolation method is quite easy and economical and is widely used. Other techniques in this context are given as follows.

3.3.1   DYNAMIC GAS DISENGAGEMENT TECHNIQUE

The method involves the measurement of liquid level or the pressure at different levels with respect to time in the reactor by gas disengagement. This technique is used either to determine the overall gas and solid hold-up in a slurry bubble column or to determine the unsteady change of overall gas hold-up and its structure in the reactor. The structure of gas hold-up refers the amount of gas bubbles separated in bubble classes according to their size where gas is distributed as a dispersed phase of bubble of different classes of size. The change of the liquid level can be identified using various techniques: X-ray, visual observation and pressure profile. The basic assumptions made in this technique are the following (Camarasa et al., 1999; Boyer, 2002):

- The dispersion is axially homogeneous when the gas feed is interrupted initially.
- No bubble–bubble interactions (i.e. bubble coalescence and break-up) during disengagement.
- The bubble classes are not interrupted by other bubble classes during its movement.
- All bubbles disengage independently.

A very simple technique to measure the gas hold-up involves the sudden closure of the gas flow (in the case of batch operation, and both liquid and gas in the case of columns operated in a continuous manner) and noting the dynamic response of the dispersion height (Sriram and Mann, 1977; Deshpande et al., 1995). Deshpande et al. (1995) reported that the log-normal distribution was found to explain the observed rate of gas disengagement. Under their conditions of operation, the bubble size showed a log-normal distribution with a single peak. The disengagement curves are fitted by two distinct curves. The first one corresponded to the sharply decreasing portion and has a slope equal to the superficial gas velocity. This portion was attributed to the disengagement of the large bubbles (Deshpande et al., 1995). The second and less steep part of the curve was attributed to the smaller bubbles entrained by the local liquid circulation. Maximum investigators used the disengagement technique as a minor tool for getting more evidence about the design parameters such as mass and heat transfer coefficients. They concluded with a unimodal size distribution for gas velocity equals to 0.02 m/s, a priori assumed a bimodal size distribution for the bubbles which are disengaged (Deshpande et al., 1995). Although this description of the gas hold-up by a two-bubble-class model is able to explain some of the observed rates of mass transfer, it cannot explain the following commonly observed phenomena in bubble columns (Deshpande et al., 1995). More details of the process are given in Majumder (2016).

3.3.2 CONDUCTIMETRY

If the liquid in a gas–liquid or gas–liquid–solid reactor is electrically conductive and presents a constant ionic concentration, the conductivity between two levels of the reactor is an increasing function of the liquid hold-up (Boyer, 2002; Uribe-Salas et al., 1994). The conductivity depends not only on the liquid hold-up but also on its distribution. A well-designed variation of the conductimetric technique was proposed by Shen and Finch (1996) to determine bubble swarm velocity in a bubble column. They developed a fast conductimeter; the bubbles disturb the conductimetry signal and the propagation celerity of this perturbation which can be interpreted as a measure of the bubble swarm velocity. Note that in this case, it is no longer necessary to know the quantitative correlation of the liquid hold-up as a function of the conductivity. Conductometric method involves the gas hold-up measurement by the classical potential theory to relate the measured conductivity to the gas hold-up. Maxwell (1892) developed that the effective conductivity of dispersion (k_{l-d}) is related to the volume fraction (ε_d) of a dispersed non-conductive phase by

$$k_{l-d} = k_l \left(\frac{1 - \varepsilon_d}{1 + 0.5\varepsilon_d} \right) \qquad (3.11)$$

Gas bubble hold-up in three-phase bubbly flow can be calculated based on this principle from the following relation:

$$\frac{1 - \varepsilon_g}{1 + 0.5\varepsilon_g} = \frac{k_{l-s-g}}{k_{l-s}} \qquad (3.12)$$

where k_{l-s} and k_{l-s-g} are the electrical conductivities of slurry and slurry–gas mixture, respectively. A schematic representation of the bubble phase hold-up measurement by electrical conductivity is shown in Figure 3.1.

The electrical conductivities of liquid–solid and liquid–solid–gas mixture can be measured by electrical conductivity meter or online conductivity probe with data acquisition system. The cell in the conductivity meter consists of two infinite and parallel plate electrodes. The electrical conductance measured by the electrodes is expressed mathematically as

$$K = \kappa \frac{A}{L_e} \qquad (3.13)$$

where K is the electrical conductance (inverse of resistance), κ is the electrical conductivity, A and L_e are the areas (m²), D is the diameter of electrode and the separation (m) of the electrodes, respectively. A/L_e is called the cell constant. Such a cell constant has been used to measure effective conductivities of dispersions. Alternating current is used to avoid polarization of the electrodes.

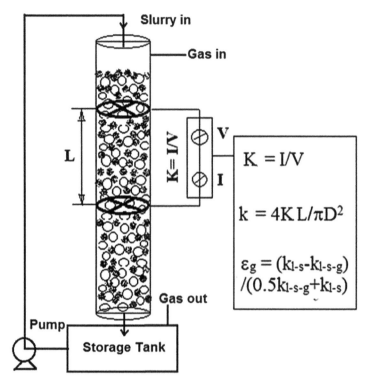

FIGURE 3.1   Schematic representation of gas hold-up measurement by electrical conductivity.

3.4   ESTIMATION OF GAS HOLD-UP IN DOWNFLOW GAS-INTERACTING SLURRY REACTOR

The overall gas hold-up in downflow gas-interacting slurry reactor can be measured by 'phase isolating method' and 'pressure drop measurement'. In this method, first, the total height of the gas–slurry mixture in the reactor is noted at steady state operation. At a certain time, the inlet and outlet of the system is closed by simultaneously quick closing valves, which will cause an immediate termination of the flow of fluids and the slurry in the column. After that, it will be allowed to settle for some time till all the gases separated from the slurry. Then after disengagement of gas, the slurry height in the column is to be noted. The difference of these two gas–slurry mixture height and only slurry height will give the overall gas phase hold-up in the slurry column reactor. The overall gas hold-up is then to be calculated as:

$$\varepsilon_g = \frac{h_m - h_{sl}}{h_m} \tag{3.14}$$

In the downflow gas-interacting slurry reactor, gas entrainment occurs due to surface braking of fluid mixture in the column by the plunging slurry jet. A minimum slurry flow rate is therefore required to move the gas as a dispersed bubbles by balancing over their (bubbles) upward buoyancy force, which depends on the bubble size and physical properties of the fluid (Majumder, 2016).

Without dynamic effect, the gas hold-up can also be assessed from the simple hydrostatic pressure as:

$$\varepsilon_g = 1 - \frac{\Delta P}{\rho_{sl} g h_m} \tag{3.15}$$

In the jet ejector-induced downflow gas-interacting slurry reactor, the sum of the forces acting on the vicinity of the jet-plunging zone can be expressed as:

$$\Delta P A_c + \rho_m g h_m A_c = \rho_m u_m (Q_{sl} + Q_g) - \rho_{sl} u_j Q_{sl} - \rho_g u_g Q_g \tag{3.16}$$

Neglecting the gas density compared to the liquid density, the above equation can be expressed as

$$\varepsilon_g = 1 - \frac{\Delta P}{\rho_{sl} g h_m} + \frac{u_j A_j}{g h_m A_c}\left(u_j\left(1 - \frac{A_j}{A_c}\right) - \frac{Q_g}{A_c}\right) \tag{3.17}$$

where Q_g is the flow rate of gas. A_c and A_j are the cross-sectional area of the reactor and jet, respectively.

3.5 EFFECT OF DIFFERENT VARIABLES ON GAS HOLD-UP IN SLURRY BUBBLE COLUMN

The gas hold-up in slurry reactor depends on the operating variables, geometric variables and thermodynamic variables. Moreover, it depends on the gas nature (molecular weight), liquid nature (aqueous, organic and mixture), liquid physical properties (density, viscosity, surface tension, vapour pressure and foaming characteristics) and solid particle nature (density and size) (Behkish et al., 2006). The details of the gas hold-up change with different operating variables in conventional slurry reactor and predicting correlations can be obtained from the literature as given in Table 1.5 in Chapter 1. The gas hold-ups of various gases (CO_2, air, N_2, He and H_2) in several organic liquids are higher than those in water (Öztürk et al., 1987; Behkish et al., 2006; Behkish et al., 2007). A maximum gas hold-up values can be obtained at a given composition mixture of liquids, such as toluene and ethanol, due to the formation of small gas bubbles. Ishibashi et al. (2001) studied the gas hold-up for three different kinds of coals in coal liquefaction reactors. They concluded that the gas hold-up, ϵ_g, increases almost linearly with superficial gas velocity in the range of 0.07–0.08 m/s and the rate of increase gradually diminishes above superficial gas velocity 0.08 m/s. They reported that the coal properties do not have significant effect on gas hold-up. For reactors with diameters larger than 0.175 m, little effect of the diameter was found on gas hold-up. Behkish et al. (2006) concluded that the total syngas hold-up increased with pressure, superficial gas velocity and temperature and decreased with increasing catalyst concentration. The total gas hold-up was also found to slightly decrease with increasing column diameter until about 0.8 m and then levelled off. The total syngas hold-ups obtained with the multiple-orifice nozzle and the spider-type gas distributor decreased with increasing orifice diameter. Moreover, the total gas hold-up obtained with the spider-type distributor was systematically greater than that with a multiple-orifice nozzle under similar conditions. They reported that the decrease of syngas hold-up with increasing catalyst concentration was due to the decrease of hold-up of small gas bubbles which disappeared at catalyst concentration of about 14 vol.%. The disappearance of small gas bubbles under such conditions was attributed to the increase of gas bubbles coalescence due to the increase of slurry viscosity. In the subsequent year, they also investigated the gas hold-up in a large-scale slurry bubble column with nitrogen and helium in Isopar-M operating in the churn-turbulent flow regime under a wide range of pressure, temperature and solid concentration. They reported that the total gas hold-up was found to increase with increasing gas density as the values for nitrogen were always greater than those for helium under the same operating conditions. Increasing pressure reduced the size of gas bubbles and subsequently increased the hold-up of small gas bubbles. They observed that increasing temperature led to the decrease of surface tension, liquid viscosity and froth stability, which resulted in the increase of the hold-up of small gas bubbles and subsequently the total gas hold-up. Increasing the superficial gas velocity slightly promotes gas bubble break-up which results in the increase of hold-up of small gas bubbles. They stated that increasing the solid concentration resulted in the increase of the hold-up of large gas bubbles due to increase in the viscosity of slurry and the coalescence tendency of gas bubbles, and decreased the froth stability. Zhang et al. (2010) studied the gas hold-up, bubble diameter and bubble rise velocity in the internal loop section of the gas–liquid–solid combined reactor which was characterized by using the double-sensor conductivity probe technique. The effects of superficial gas velocity, external slurry circulation or solid loading on the hydrodynamics in the internal loop section of the slurry reactor were investigated. They concluded that the gas hold-up within the draft tube is greatly affected by its geometrical configuration, and is much higher than that in the corresponding section of the annular region. They reported that operational conditions, such as superficial gas velocity, external slurry circulation velocity and solid loading, have an impact on the hydrodynamics in the internal loop section. Local gas hold-ups in all regions, axial distribution of section-averaged gas hold-ups in the inside of draft tube and the annular region, overall gas hold-up in the internal loop section, local mean bubble diameters and local bubble rise velocities in all regions increase with an increase of superficial gas velocity. They observed that the effects of external slurry circulation velocity and solid loading on the hydrodynamics are insignificant compared with that of superficial gas velocity. However, they stated that local, section-averaged and overall gas hold-ups have a slight increase while local mean bubble diameter and averaged local bubble rise velocity decrease when external slurry circulation or solid loading increases. Götz et al. (2016) investigated the parameters influencing the hydrodynamics of a slurry bubble column reactor operated at elevated temperatures up to 573 K with heat transfer oils and ionic liquids. They observed that gas spargers with small gas velocities per hole cause smaller bubbles and higher gas hold-ups. They reported that the use of a perforated plate with small holes (100 μm) and a defined pitch of holes of 1×10^{-3} m in combination with non-coalescing media enables the formation of small bubbles ($1–2 \times 10^{-3}$ m) and a large gas–liquid interfacial area (up to 700 m^{-1}). They concluded that the gas density was found to have a small effect on the gas hold-up in their investigated range (mostly homogeneous regime). The influence of the liquid properties is different for different flow regimes. They reported the following observations from their study: (i) in the homogeneous regime, an increasing viscosity has no effect or even a positive effect on the gas hold-up, (ii) in the homogeneous regime, the surface tension is the most important liquid property, (iii) an increase in viscosity and surface tension destabilizes the homogeneous regime, (iv) the hydrodynamics of the investigated heat transfer oils is significantly influenced by surfactants, (v) the use of ionic liquids leads to bubble coalescence and consequently to low gas hold-ups and to a desta-bilization of the homogeneous regime, (vi) the hydrodynamic behaviour of ionic liquids is ruled by their viscosity at low temperatures (e.g. small $u_{G,trans}$) and the surface tension at elevated temperatures (>373 K), (vii) all the investigated ionic liquids show a similar hydrodynamic behaviour, (viii) particles increase the primary bubble size formed at the gas sparger, (ix) small concentrations of small particles ($\ll 100$ µm) can reduce bubble coalescence, otherwise particles enhance bubble coalescence and (x) an increase in density difference between solid and liquid density (e.g. by increasing the particle density) reduces the gas hold-up. Abdulrahman (2016) investigated the overall gas hold-up experimentally for helium gas at 90°C injected through a slurry of water at 22°C and alumina solid parti-cles in a slurry bubble column reactor. His work examines the effects of superficial gas velocity, static liquid height, solid particles concentration and solid particle diameter on the overall gas hold-up of the slurry bubble column reactor. These effects are expressed in the forms of empirical equations. He observed that the overall gas hold-up increases by increasing the superficial gas velocity with a higher rate of increase at lower superfi-cial gas velocity. In addition, the overall gas hold-up decreases by increasing the static liquid height and/or the solid concentration at any given superficial gas velocity. He concluded that at a higher solid concen-tration, the changing rate of the overall gas hold-up with the superficial gas velocity and/or the solid concentration is lower. He also reported that the effect of the solid particle diameter on overall gas hold-up is negli-gible. The effect of solid particles, including magnesium hydroxide, calcium hydroxide, iron oxide, calcium carbonate and carbon particles, at various concentrations in slurry reactors was reported to increase the gas hold-up and gas–liquid interfacial area at low concentrations (<5 vol.%) (Behkish et al., 2006). The increase of solid particle size, on the other hand, was found to decrease the gas hold-up. Behkish et al. (2006) also reported that the effect of solid particles on the gas hold-up should account not only for the solid concentration but also for particle nature, size and density, which might significantly affect the gas hold-up and subsequently the gas–liquid interfacial area.

Gas distributor is one of the important parts of the two- or three-phase reactors which may regulate the gas distribution in the reactor. There are various types of gas distributor such as spider, perforated plate, sintered plate, nozzles, and so forth, which significantly differ in their size and number of holes in the distributor plate. Commonly used gas distributors are perforated plates, porous plates, sintered plates, single-orifice nozzles, multiple-orifice nozzles, rings, spider types, bubble caps and injector and ejector types (Behkish et al., 2006). The performance of the gas distributor depends on the opening size, number of openings, sparger positioning and nozzles position/orientation in the reactor. The effect of gas sparger on the gas hold-up yet not fully understood. Gas distributor has a minimal effect or almost no effect on the bubble sizes and gas hold-up if the orifice diameters are > 1–2 mm (Wilkinson et al., 1992; Jordan and Schumpe, 2001). The gas hold-up depends on the break-up and coalescence of the gas bubbles in the column. The distributor of smaller pore diameters (porous plates) generates smaller gas bubbles compared to perforated plates. Single-orifice nozzles, with diameters usually greater than 0.001–0.002 m, generated large gas bubbles, even at very low superficial gas velocity, indicating a heterogeneous bubble size distribution (Akita and Yoshida, 1974; Camarasa et al., 1999). Therefore, it can be concluded that *gas hold-up* is inversely proportional to the orifice diameter under the negligible coalescence of the bubble. In this regard, the work of Schügerl et al. (1977) is notable. They studied the effect of the gas distributors on gas hold-up to obtain a non-coalescing system. Furthermore, Schügerl et al. (1977) showed that in a coalescing system, the effect of gas distributor on the gas hold-up values was not significant, confirming that in a non-coalescing system. Therefore, it is worthy to say that the bubble size distribution is controlled by the gas distributor and if the gas–liquid system in a slurry bubble column reactor is non-coalescing, one can expect that the bubble size distribution and subsequently the gas hold-up would be strongly dependent on the gas distributor design.

It is also important to say that the hydrodynamics of slurry bubble column reactors is strongly dependent on the column geometry. For longer column, the gas is distributed axially with different mechanisms. In the gas distributor region, the gas hold-up depends on the design of the distributor. In the bulk region, the gas hold-up is controlled by the liquid/slurry circulation, whereas in the top region of the column, the gas hold-up is large due to the formation of a layer of froth above the liquid/slurry bed. In general, the gas hold-up is the sum of the hold-ups in the three regions; however, if the column is long enough, the influence of the first and third regions on the gas hold-up will be insignificant and thus the gas hold-up will be close to the values measured in the bulk region (Wilkinson et al., 1992; Behkish et al., 2002, 2006). The ratio of height of the reactor to its diameter would therefore affect the gas hold-up. In the literature, it is reported that typically no obvious change in the gas hold-up was observed when column height to diameter ratios were >5–6 (Wilkinson et al., 1992; Behkish et al., 2002, 2006). Furthermore, the gas hold-up was found to decrease with column diameter (Sarrafi et al., 1999) due to a reduction in the hold-up of large gas bubbles, a change in the liquid backmixing and a reduction of the foaming ability of the liquid/slurry. Similarly, many investigators have reported that the gas hold-up would level off when column diameters are ≥0.15 m (Akita and Yoshida, 1973; Kastanek et al., 1984; Wilkinson et al., 1992; Behkish et al., 2002; Shah et al., 2012). However, Koide et al. (1984) observed a small influence of the column diameters on the gas hold-up. They suggested that the difference was negligible. In addition, they pointed out that if larger gas bubbles were formed in larger columns, a relatively smaller total gas hold-up would be expected. Behkish et al. (2006) reported that the gas hold-up continues to slightly decrease at column diameter >0.15 m and it will result to reach an asymptote value based on experimental conditions.

3.6   EFFECT OF OPERATING VARIABLES ON GAS HOLD-UP IN A DOWNFLOW GAS-INTERACTING SLURRY REACTOR

The overall gas hold-up is found to decrease with increase in separator pressure in the downflow ejector-induced slurry bubble column. As the separator pressure increases, the bubbles face more resistance to move downward and to gas entrainment (Sivaiah et al., 2012). The overall gas hold-up is higher at the higher concentration of solids. The bubbles inside the column face more resistance to move in the downward direction due to increase of the viscosity of the solution. Thus, the residence time of the gas bubbles inside the column increases, which in turn increases the overall gas hold-up. The separator pressure does not vary significantly with different concentrations of the slurry. The gas hold-up changed with separator pressure can be related as $\varepsilon_g = a - bP_s$. The coefficients 'a' and 'b' depend on the slurry concentration. Some typical values are shown in Table 3.1.

TABLE 3.1   Typical Values of a and b.

Slurry concentration	Coefficient 'a'	Coefficient 'b'
0.1	1.0924	1.4267
0.5	1.1105	1.452
1	1.1924	1.5701

The superficial slurry velocity is a dominant factor that affects the gas hold-up. Generally, the gas hold-up increases with an increase in the superficial liquid slurry velocity (Sivaiah et al., 2012).

The average gas hold-up increases almost linearly with increasing superficial slurry velocity. As the slurry velocity increases, the momentum transfer increases which enhances the entrainment of the gas in the column. High-velocity liquid jet producing the high kinetic energy results in more entrainment of the gas. Moreover, as the slurry concentration increases, the resistance of the fluid increases which results decrease in both gas entrainment and gas hold-up (Sivaiah and Majumder, 2013). It is observed that the gas hold-up increased with increase in slurry velocity as well as gas velocity. With the increase in superficial slurry velocity through the nozzle, the jet momentum increases, the secondary air disperses into discrete bubbles which, in turn, increase the gas hold-up (Sivaiah and Majumder, 2013). Generally, the gas hold-up increases with an increase in the superficial gas velocity. The average gas hold-up increases almost linearly with increasing superficial gas velocity in the homogeneous regime. In the homogeneous system, the population of the bubble is higher due to formation of more uniform finer bubbles, whereas in heterogeneous regime, bubble size is not uniform (Sivaiah and Majumder, 2013). The bigger size bubbles immediately get released due to their higher buoyancy. It is observed that the gas hold-up is higher when the outlet superficial gas velocity is fully closed than it is controlled. When the separator outlet is closed, the entrained gases face more resistance to move in downward direction and spend more time in the column which results in more mixing, bubble break-up to form uniform bubble, and hence holding more gas in the column. With increase in particle diameter, the break-up of the bubbles increases in the column and enhances to increase the number density of the smaller bubbles (Sivaiah and Majumder, 2013).

These smaller bubbles with lower rising velocity lead to form higher residence time in the column and consequently result the higher gas hold-up (Sivaiah and Majumder, 2013). It is observed that the gas hold-up decreases with increasing slurry concentration. This is due to the fact that with increase in slurry concentration, bubble faces more resistance to move downward direction because of their buoyancy effect and viscous liquid medium (Sivaiah and Majumder, 2013). A comparative result of gas hold-up obtained by different authors over a range of superficial gas and liquid velocities and type of solid used in downflow gas-interacting slurry reactor is shown in Table 3.2. Higher gas hold-up was observed in the downflow gas-interacting slurry reactor compared to others due to higher residence time of the bubbles. In this downflow gas-interacting slurry reactor, the range of gas hold-up obtained with the air–water–glycerol–solid system is 0.01–0.61 (Sivaiah and Majumder, 2013).

3.7   ANALYSIS OF GAS HOLD-UP

3.7.1   SLIP VELOCITY MODEL

Gas hold-up is a design parameter which affects interracial area between the phases. Various models are available to analyse the dispersed phase hold-up in multiphase contactors (Majumder, 2016). The models are suitable as per operating range, the geometry and physical property range of the systems. In this context, the slip velocity model is developed for the application in the three-phase flow in the downflow gas-interacting slurry reactor.

In the downflow system, the slip velocity can be calculated by the resultant velocity of bubbles relative to downward slurry velocity as

$$u_s = \pm \left\{ u_b - \frac{u_{sl}}{1 - \varepsilon_g} \right\} \qquad (3.18)$$

TABLE 3.2 Literature Results of Range of Gas Hold-up of Downflow Gas-Interacting Slurry Reactor.

Reference	Type of fluid used	Column diameter (m)	Particle diameter (μm)	Type of gas distributor	Operating range (m/s)		Range of gas hold-up (ε_g)
					Liquid	Gas	
Shah et al. (1983)	Air/water, sodium chloride, alcohol	0.075	–	Ring-type distributor	0.20–0.32	0.0006–0.02	0.029–0.320
Ohkawa et al. (1986)	Air/water	0.02–0.026	–	Nozzle	0.05–0.20	0.05–0.20	0.01–0.40
Bando et al. (1988)	Air/water	0.07	–	Perforated sparger	0.01–0.20	0.01–0.10	0.01–0.345
Yamagiwa et al. (1990)	Air/water	0.034–0.07	–	Jet	0.40–0.91	0.10–0.50	0.15–0.40
Wachi et al. (1991)	–	0.053	40	Multi-orifice	–	–	0.58
Marchese et al. (1992)	Air/water	0.038	75	–	–	–	0.2–0.6
Briens et al. (1992)	Air/water	0.095	–	–	0.20–0.50	0.01–0.10	0.15–0.40
Douek et al. (1997)	Air/water	0.08	484	Sparger	0.37–0.50	0.01–0.04	0.04–0.08
Sivaiah and Majumder (2013)	Air/water, glycerol	0.05	2.21–96	Nozzle	0.04–0.14	$0.85 \times 10^{-2} – 2.55 \times 10^{-2}$	0.01–0.61

The bubble rise velocity can be calculated by the correlation proposed by Evans et al. (1992) as

$$u_b = 0.496 \left(\frac{gd_c}{2} \right)^{0.5}$$ (3.19)

Majumder et al. (2006) enunciated the slip velocity for two-phase flow in downflow reactor as a function of gas hold-up. The slip velocity for churn-turbulent regime (Lapidus and Elgin, 1957) is expressed as

$$u_s = u_b f(\varepsilon_g)$$ (3.20)

where u_b is the bubble terminal rise velocity and $f(\varepsilon_g)$ shows the effect of the interaction of neighbouring bubbles within the column. According to Hills (1976), the rise velocity of bubble is 1.0 m/s and the interaction term of the neighbouring bubbles is proportional to 1.72 power of the gas hold-up for $u_l \leq 0.3$ m/s, whereas Clark and Flemmer (1985) reported the order of the variation of the interaction term is 0.72 and bubble rise velocity is 0.25 m/s. They represented the slip velocity by the following equation:

$$u_s = u_b (1 - \varepsilon_g)^\alpha$$ (3.21)

They found best fit value $u_b = 0.25$ m/s and $\alpha = 0.702$ with their experimental data. Several other forms of slip velocity models developed by various authors are shown in Table 3.3.

Different authors analysed their experimental data by different functions of gas hold-up to express the slip velocity. The models are applicable for upward bubble flow. In case of downflow gas-interacting slurry reactor, it was observed that slip velocity is not only a function of gas hold-up but also a function of the slurry concentration (Sivaiah et al., 2012). In the case of downflow system, the slip velocity may be positive or negative. The positive value of slip velocity indicates that the rise velocity of bubbles is higher than the slurry liquid velocity and bubbles cannot be carried downward along with the downflow slurry liquid. Otherwise, the bubbles are carried downward by this slip velocity. The bubbles try to move downward along the liquid slurry when the slurry velocity is greater than bubble rise velocity as described in our study (Sivaiah et al., 2012). Moreover, from the present experimental data, it is seen that for this downward slurry

TABLE 3.3 Different Slip Velocity Models Developed by Different Authors in Multiphase Systems.

Models developed by	Model equations	Parameters
Marrucci (1965)	$u_s = u_b(1-\varepsilon_g)/(p-q\varepsilon_g^r)$	u_b=undefined; p=1.0; q=1.0; r=5/3
Lockett and Kirkpatrick (1975)	$u_s = u_b(1-\varepsilon_g)^k(1+l\varepsilon_g^m)$	u_b=undefined; k=1.39; l=2.55; m=3
Hills (1976)	$u_s = \beta + \gamma\varepsilon_g^\delta$	β=0.24; γ=4.0; δ=1.72
		β=0.23; γ=4.0; δ=1.73
Shah et al. (1983)	$u_s = u_b(1-\varepsilon_g)^{n-1}$	u_b=28.17; n=2.74
		u_b=undefined; n=2.39
Clark and Flemmer (1985)	$u_s = u_b[\varepsilon_g(1-\varepsilon_g)]^n$	u_b=22.81; n=−1
Jamialahmadi and Muller-Steinhagen (1993)	$u_s = \beta + \gamma\varepsilon_g^\delta$	β=0.24; γ=4.0; δ=1.72
		β=0.23; γ=4.0; δ=1.73
Zahradnik et al. (1997)	$u_s = u_b(1-\varepsilon_g^\varphi)^\psi$	u_b=1.0; φ=1.0; ψ=−1
Sarrafi et al. (1999)	$u_s = u_b(1-\varepsilon_g)^{n-1}$	u_b=28.17; n=2.74
		u_b=undefined; n=2.39
Majumder et al. (2006)	$u_s\mu_l/\sigma_l = \lambda D_R^s\varepsilon_g^t$	u_b=1.10–6.84; s=4.26; t=−0.64; λ=0.015
	$u_b = \lambda(\sigma_l/\mu_l)(D_R)^s$	
Sivaiah et al. (2012)	$u_s = a\ln(1-\varepsilon_g)+b$	a=0.38–0.40; b=0.232–0.219

bubble flow, the experimental value of effective slip velocity does not fit well to the other models. Sivaiah et al. (2012) developed the following correlation for slip velocity in the ejector-induced gas-interacting slurry reactor as:

$$u_s = a\ln\left(1-\varepsilon_g\right) + b \tag{3.22}$$

where

$$a = 0.3901 + 0.0051w + 0.0075w^2 \tag{3.23}$$

$$b = 0.2403 - 0.003w - 0.0005w^2 \tag{3.24}$$

At zero slip velocity, the gas hold-up is represented as critical gas hold-up which is yielded from Equation 3.22 as follows:

$$\varepsilon_{g,c} = 1 - e^{(-b/a)} \tag{3.25}$$

The values of critical gas hold-up are 0.4590 at $w=0.1$ and 0.4446 at $w=1.0$. The critical gas hold-up decreases with increase in slurry concentration due to increase in apparent viscosity with the slurry concentration (Sivaiah et al., 2012).

3.7.2   ANALYSIS OF GAS HOLD-UP BY DRIFT FLUX MODEL

The drift flux model is generally used to analyse the experimental hold-up data for two-phase vertical system (Zuber and Findlay, 1965). This can be extended to three-phase flow. The model has been developed for the application of three-phase flow in the downflow gas-interacting slurry reactor. It is reported in our study published in 2013 (Sivaiah and Majumder, 2013). Zuber and Findlay represented the model based on superficial gas velocity, gas hold-up and mixture velocity (gas–liquid) as

$$\frac{u_g}{\varepsilon_g} = c_o(u_g + u_{sl}) + u_d \tag{3.26}$$

where u_d is the drift velocity of gas phase and c_o is the distribution parameter. The value of c_o depends on the difference between hold-up profiles and velocity across the column. In Equation 3.26, if the values of c_o and u_d are equal to 1 and 0, respectively, then the flow is described as homogeneous flow. If the value of u_d is equal to 0 and c_o is not equal to 1, then the drift flux relationships are termed as slip equations. The drift velocity developed for upward bubbly flow (Darton and Harrison, 1975; Kawanishi et al., 1990) can be expressed as

$$u_d = \sqrt{2}\left(\frac{g\sigma_l(\rho_l - \rho_g)}{\rho_l^2}\right)^{1/4} \tag{3.27}$$

For downflow gas-interacting slurry bubble column, the distribution parameter (c_o) can be evaluated as follows (Sivaiah and Majumder, 2013)

$$c_o = \frac{\left(\dfrac{u_g}{\varepsilon_g}\right) + \sqrt{2}\left[\dfrac{g\sigma_l\left(\rho_l - \rho_g\right)}{\rho_l^2}\right]^{1/4}}{u_g + u_{sl}} \qquad (3.28)$$

It is found that the distribution parameter (c_o) is a function of particle diameter, physical and dynamic variables of the system. Correlations have also been made by dimensional analysis to obtain the values of c_o and u_d within the experimental operating ranges of superficial slurry velocity 0.04–0.14 m/s and the superficial gas velocity 0.85×10^{-2} to 2.55×10^{-2} m/s. It can be expressed as (Sivaiah and Majumder, 2013)

$$c_o = 81.712\, Re_{sl}^{-1.366}\, Re_g^{0.075}\, D_r^{-0.014}\, M_o^{-0.328} \qquad (3.29)$$

with correlation coefficient $(R^2) = 0.981$ and standard error $(SE) = 0.076$. The ranges of the operating variables of the correlation Equation 3.29 are $8.73 \times 10^2 < Re_{sl} < 49.03 \times 10^2$, $27.82 < Re_g < 83.47$, $0.52 \times 10^3 < D_r < 22.68 \times 10^3$ and $1.04 \times 10^{-10} < Mo < 6.27 \times 10^{-10}$. A comparison between the present experimental results of the drift velocity (u_d) and distribution parameter (c_o) with the other published three-phase works are compiled in Table 3.4. The typical range of average value of the drift velocity (u_d) and distribution parameter (c_o) for glycerol and aluminium oxide slurry of concentration 0.1 to 1.0% is -0.2246 to -0.2226 and 4.379–5.156, respectively. In case of gas-interacting downflow slurry reactor, the values of distribution parameter are higher at higher slurry concentration (Sivaiah and Majumder, 2013). This is due the reason that with increase in slurry concentration, the particle mobility and velocity are reduced for the higher friction losses between the particles and the viscous liquid medium. Consequently, the break-up of the bubbles inside the column is more at higher slurry concentrations. The negative value of the drift velocity refers to the downflow system (Sivaiah and Majumder, 2013).

3.7.3   ANALYSIS OF GAS HOLD-UP BY LOCKHART–MARTINELLI MODEL

Lockhart–Martinelli (Lockhart–Martinelli, 1949) correlation can be modified to assess the gas hold-up in three-phase flow though the model developed for two-phase flow. In case of three-phase flow, the liquid phase

TABLE 3.4 Comparison of Drift Velocity (u_d) and Distribution Parameter (c_o) for Up- and Downflow Three-Phase Flow.

Mode of operation	Drift velocity (u_d) (m/s)	Distribution parameter (c_o)	Flow regime	Reference
Upflow	$\varepsilon_g(1-\varepsilon_g)$ $\times(u_g/\varepsilon_g - u_l/\varepsilon_l)$	—	Bubbly flow	Darton and Harrison (1975)
Downflow	0.20	1.0	Bubbly flow	Bando et al. (1990)
Upflow	$V_d = V_b \phi_l^m \delta_l^{m_2}$	$1/(1-\varepsilon_{so})$	Bubbly flow	Chen and Fan (1990)
Upflow	$V_d = kV_b$	1.01	Bubbly flow	Douek et al. (1994)
Upflow	0.21	1.15	Bubbly flow	Douek et al. (1997)
Downflow	0.41	1.13	Bubbly flow	Douek et al. (1997)
Downflow	$\sqrt{2}\left(\dfrac{g\sigma_l(\rho_{sl}-\rho_g)}{\rho_{sl}^2}\right)^{1/4}$	$81.712\times\mathrm{Re}_{sl}^{-1.366}$ $\times\mathrm{Re}_g^{0.075}\times D_r^{-0.014}$ $\times M_o^{-0.328}$	Bubbly flow	Sivaiah and Majumder (2013)

could be considered as a slurry phase. The gas hold-up can be analysed by the modifications of Lockhart–Martinelli parameter.

$$X = \phi_g / \phi_{sl} = \sqrt{\Delta P_{fosl} / \Delta P_{fog}} \qquad (3.30)$$

Butterworth (1975) presented the Lockhart–Martinelli relation for gas hold-up based on quality (x) and fluid properties such as density (ρ) and viscosity (μ). An attempt has been made to modify the correlation in term of quality (x) for downflow gas-interacting slurry reactor. The quality is equal to the ratio of the mass flow rate of gas $(\dot{m}_g)$ and the mass flow rate of gas–liquid–solid mixture $(\dot{m}_m)$.

$$x = \frac{\rho_g Q_g}{\left(\rho_g Q_g + \rho_{sl} Q_{sl}\right)} \qquad (3.31)$$

The volumetric flow rate of the gas (Q_g) in the downflow flow system can be calculate from the volumetric flow rate of the solid–liquid slurry (Q_{sl}) and the gas hold-up (ε_g) as

$$Q_g = \frac{\varepsilon_g Q_{sl}}{\left(1 - \varepsilon_g\right)} \qquad (3.32)$$

By substituting Equation 3.32 in Equation 3.31, the gas hold-up can be expressed as

$$\varepsilon_g = \frac{1}{\left(1 + \dfrac{\rho_g}{\rho_{sl}} \dfrac{(1-x)}{x}\right)} \qquad (3.33)$$

The quality (x) has been correlated by Sivaiah and Majumder (2012) with the physical and dynamic variables of the system which can be expressed as

$$\frac{1-x}{x} = 1.007 \times 10^{10} \ Re_{slb}^{-0.249} \ Re_g^{-1.205} \ D_r^{-0.136} X^{-2.253} \qquad (3.34)$$

From Equations 3.33 and 3.34, one gets

$$\varepsilon_g = \frac{1}{\left(1 + \left(\dfrac{\rho_g}{\rho_{sl}}\right)\left(1.007 \times 10^{10} \ Re_{slb}^{-0.249} \ Re_g^{-1.205} \ D_r^{-0.136} X^{-2.253}\right)\right)} \qquad (3.35)$$

It is seen that the developed correlation fits well within the range of variables: $22.27 < Re_{slb} < 558.2$, $27.82 < Re_g < 83.47$ and $0.52 \times 10^3 < D_r < 22.68 \times 10^3$.

3.7.4 OTHER CORRELATION MODELS

Other different forms of correlation model used for prediction of gas hold-up in the slurry reactor are given as follows:

3.7.4.1 HOMOGENEOUS FLOW MODEL

In this model, it is assumed that the fluids behave like a homogeneous mixture. The velocity and the density of the mixture are considered constant across the tube and the gas hold-up is predicted from the flow rate of the fluids entering the system by the following relationship

$$\varepsilon_g = \frac{Q_g}{Q_g + Q_l} \tag{3.36}$$

Homogeneous model gives reasonable prediction of the gas hold-up if pressure losses are negligible (Isbin et al., 1957). This model leads to larger errors when pressure losses are significant.

3.7.4.2 VARIABLE-DENSITY MODEL

Bankoff (1960) modified the homogeneous flow model to consider the effect of radial non-uniformity and concentration. He suggested that although there is no slip between the gas and the liquid in the direction of flow, gradient in bubble concentration exists in the axial direction. By assuming a power law distribution for both the velocity and the concentration profiles in the radial direction, Bankoff proposed the following relationship between the liquid hold-up and the input flow rates

$$\varepsilon_l = K \frac{Q_l}{Q_l + Q_g} \tag{3.37}$$

where K is the Bankoff factor. Although the Bankoff model was developed for vertical flow, the model can be used to analyse the gas hold-up in the downflow gas-interacting slurry reactor. In this case, the Bankoff factor depends on the slurry concentration and the particle size.

3.7.4.3 MOMENTUM EXCHANGE MODEL

In the gas-interacting slurry reactor, the sum of the frictional and hydro-static losses in one phase equals the sum of the same losses in the other phase; there would be a rapid exchange of momentum between the phases to maintain the equilibrium condition. On the basis of this statement, the gas hold-up in the gas-interacting slurry reactor can be analysed by the model developed by Levy (1966). Levy (1966) proposed the momentum exchange model on this condition. According to this model, the gas hold-up is can be expressed as

$$\frac{1}{\varepsilon_g (1+R_m)^2} + \frac{\rho_l}{2(1+R_m)\rho_g} = \frac{1}{2(1+R_m)^2(1-\varepsilon_g)^2} \tag{3.38}$$

where R_m is the gas/liquid mass flow ratio. The Levy model gives reason-able agreement with experimental data in horizontal tubes for high void volumes. The model can be used for slurry reactor provided the consider-ation of fluid properties change with the slurry concentration.

3.7.4.4 BAROCZY CORRELATION

According to Baroczy (1963), the void fraction data can be correlated graphically in the form

$$1-\varepsilon_g = f_1(X,\Lambda) \tag{3.39}$$

where

$$\Lambda \propto \left(\frac{\rho_g}{\rho_l}\right)\left(\frac{\mu_l}{\mu_g}\right)^a \tag{3.40}$$

Equation 3.36 can also be represented by

$$\frac{1-\varepsilon_g}{\varepsilon_g} = f_2(\frac{1-x}{x},\Lambda) \tag{3.41}$$

where X can be expressed as

$$X \propto \left(\frac{1-x}{x}\right)^{b} \Lambda^{c}$$
(3.42)

They reported that the function f_2 is approximated over most of the range of x and Λ by

$$\frac{1-\varepsilon_g}{\varepsilon_g} \propto \left(\frac{1-x}{x}\right)^{b} \Lambda^{c}$$
(3.43)

Combining Equations 3.40 and 3.43, one can apply the correlation model for the gas-interacting downflow slurry bubble column and can be repre-sented as a general form as

$$\frac{1-\varepsilon_g}{\varepsilon_g} = \lambda\left(\frac{1-x}{x}\right)^{p}\left(\frac{\rho_g}{\rho_l}\right)^{q}\left(\frac{\mu_l}{\mu_g}\right)^{r}$$
(3.44)

where λ is a constant, explicitly in terms of the void fraction. The coeffi-cient λ and indices p–r can be obtained from the experimental data. Butter-worth (1975) represented the general form of correlations as a function of quality, density ratio and viscosity ratio in the same form mentioned in Equation 3.44. Explicitly in terms of the void fraction Equation 3.44 can be expressed as

$$\varepsilon_g = \frac{1}{1+\lambda\left(\frac{1-x}{x}\right)^{p}\left(\frac{\rho_g}{\rho_l}\right)^{q}\left(\frac{\mu_l}{\mu_g}\right)^{r}}$$
(3.45)

The gas hold-up can also be expressed in terms of slip ratio where the slip ratio can be expressed as

$$\frac{u_g}{u_l} = \left(\frac{x}{1-x}\right)\left(\frac{\rho_l}{\rho_g}\right)\left(\frac{1-\varepsilon_g}{\varepsilon_g}\right)$$
(3.46)

3.7.4.5   KAWASE AND MOO-YOUNG MODEL

The model is based on the momentum change due to liquid motion in the column (Kawase and Moo-Young, 1987; Majumder, 2016). The equation of motion for the liquid phase in a bubble column may be written as:

$$-\frac{1}{r}\frac{d}{dr}(r\tau) = -(\overline{\varepsilon}_g - \varepsilon_g)\rho_l g \tag{3.47}$$

$$\tau = -(v_l + v_{t,l})\rho_l \frac{du}{dr} \tag{3.48}$$

$$\varepsilon_g = \overline{\varepsilon}_g\left(\frac{2+n}{n}\right)\left[1 - \left(\frac{r}{R}\right)^n\right] \tag{3.49}$$

The flow behaviour of the liquid can be represented by a power-law model as

$$\tau = K\dot{\gamma}^n \tag{3.50}$$

At turbulent condition, the molecular kinematic viscosity, v_l, is insignificant compared with the turbulent kinematic viscosity, v_t. The gas hold-up profile in the literature indicates a value for n as 2 (Ueyama and Miyauchi, 1979). Therefore, a general solution of Equation 3.48 is written as

$$u = \frac{\overline{\varepsilon}_g g}{v_{l,t}}\left(\frac{1}{8}\frac{r^4}{R^2} - \frac{1}{4}r^2\right) + A_1 \ln(r) + A_2 \tag{3.51}$$

The constants of integration A_1 and A_2 can be resolute using appropriate boundary conditions. In case of upward flow of bubble in bubble column, the liquid rises with the bubbles in the centre of the column and flows downward in the outer annular region. The liquid velocity is maximum at the centre of the column and decreases in a radial direction. At high Reynolds numbers, a transition point at which the time average velocity is zero occurs at around $r/R=0.7$, and large velocity, the magnitude of which is 50–100% of the velocity at the column axis, is found very close to the column wall (Kawase and Moo-Young 1987). On the basis of these observations, they have formulated the following boundary conditions:

$$u = u_{l0} \quad at \quad r = 0 \tag{3.52}$$

$$u = 0 \quad at \quad r = 0.7R \tag{3.53}$$

$$u = (-0.5 \sim -1.0)u_{l0} \quad at \quad r = R \tag{3.54}$$

They also pointed out that profile in non-Newtonian fluids at high Reynolds numbers is similar to that in Newtonian fluids described above. The constants of Equation 3.51 are determined by using only boundary condition Equation 3.52 and solution can be written as

$$u = \frac{\bar{\varepsilon}_g g}{V_{1,t}} \left(\frac{1}{8} \frac{r^4}{R^2} - \frac{1}{4} r^2 \right) + u_{10} \tag{3.55}$$

It is noted that in order to represent a liquid, velocity profile in the core region of the column which may characterize the mixing in a whole bubble column. Substitution of Equation 3.54 into Equation 3.55 yields an expression for the average gas hold-up as

$$\bar{\varepsilon}_g = 43 \frac{V_{1,t} u_{10}}{d_c^2 g} \tag{3.56}$$

The turbulent kinematic viscosity may vary with the distance from the wall. A characteristic kinematic viscosity can be defined in a bubble column on the basis of the Prandtl mixing length theory (Prandtl, 1942)

$$\bar{V}_{1,t} = \frac{\chi^2}{2} u_{10} R \tag{3.57}$$

$$\chi = 0.4n \tag{3.58}$$

The model can also be applied in the gas-interacting downflow slurry bubble column provided that the slurry is considered as a continuous medium as a liquid.

3.7.5   GENERAL CORRELATION MODEL

As per different experimental condition, the gas hold-up in the downflow gas-aided slurry reactor depends on the slurry velocity (u_{sl}), gas velocity (u_g), column diameter (d_c), particle diameter (d_p), slurry density (ρ_{sl}), slurry viscosity (μ_{sl}) and surface tension of the liquid (σ_l). Hence, a general correlation can be made to correlate the gas hold-up (ε_g) by the dimensional analysis with these variables. The gas hold-up (ε_g) as a function of all these variables can be written as

$$\varepsilon_g = f(u_{sl}, u_g, \mu_{sl}, \rho_{sl}, d_c, h_m, d_p, \sigma_l, g) \tag{3.59}$$

After dimensional analysis, the variables can be related based on the different significant dimensionless groups which can be expressed as

$$\varepsilon_g = \lambda \left(\frac{u_{sl}}{u_g}\right)^{a_1} \left(\frac{d_c u_{sl} \rho_{sl}}{\mu_{sl}}\right)^{b_1} \left(\frac{u_{sl}^2}{gd_c}\right)^{c_1} \left(\frac{h_m}{d_p}\right)^{d1} \tag{3.60}$$

The coefficients, λ, a_1, b_1, c_1 and d_1 can be obtained by multiple linear regression analysis by any software. The developed correlation for down-flow gas-aided slurry reactor after multiple linear regression analysis can be expressed by Equation 3.61 with the correlation coefficient of 0.987 and *SE* of 0.098.

$$\varepsilon_g = 1.05 \times 10^{-6} \left(\frac{u_{sl}}{u_g}\right)^{-0.118} \left(\frac{d_c u_{sl} \rho_{sl}}{\mu_{sl}}\right)^{2.211} \left(\frac{u_{sl}^2}{gd_c}\right)^{-0.104} \left(\frac{h_m}{d_p}\right)^{-0.519} \tag{3.61}$$

According to different reported correlations, distinguishing the best correlation is the most important factor for scaling up the gas-interacting slurry reactor. Behkish et al. (2006) developed two novel correlations based on literature data of gas hold-up for different gases in various liquids and slurries using bubble and slurry bubble column reactors operating under wide ranges of conditions in different size reactors provided with a variety of gas spargers. One is for the total gas hold-up and the other for the hold-up of large gas bubbles. As per his developed correlations, the total gas hold-up correlation is capable of predicting the experimental data within an absolute average relative error and standard deviation of 20%, whereas the correlation of the hold-up of large bubbles is capable of predicting the experimental values within error and deviation of about 25 and 27%, respectively. Mowla et al. (2016) performed a meta-analysis on experimental data of average gas hold-up in gas–solid–liquid reactors published in numerous studies over the past 40 years. The analysis succeeded in developing statistically significant and acceptably accurate correlations for the prediction of average gas hold-up in systems involving spherical particles and pure water as the liquid phase. They have considered average bubble size and gas density to improve the correlations for the prediction of gas hold-up in high pressure systems. However, the analysis illustrated the limitations of a simple functional to approximate the outcome of complex momentum transfer in a three-phase system. Some important correlations in gas–slurry system reported by various authors are presented in Table 3.5.

TABLE 3.5 Gas Hold-up Correlations Available in the Literature for Gas–Liquid–Solid System.

Author	System studied	Range of Operating Variables	Available correlation(s)
Ziganshin and Ermakova (1970)	Air/water, aqueous glycerine solution/glass spheres	d_c: 100 and 200 mm i.d.	$\varepsilon_g = 1 - \dfrac{1}{1+1000(\mu/\rho)_{sl}^{0.5}u_g}$
Vail et al. (1970)	Air/water/glass beads, spherical aluminosilicate catalyst, spherical alumino–cobalt–molybdenum catalyst	d_c: 146 mm i.d.	$\varepsilon_g = k(1-\varepsilon_s)^m \left(\dfrac{u_g}{u_{sl}}\right)^n$; k, m and n are adjustable parameters for different particles, depend on the size of the particle. For 0.73 mm glass beads, the respective values are 0.1026, 0.780 and 2.09; for 0.77 mm Al–Si catalyst, values are 0.108, 0.63 and 3.01; and for 0.74 mm Co–Mo catalyst, those are 0.0526, 0.670 and 1.69.
Kumar et al. (1976)	Air/H$_2$O, glycerol, kerosene	P: atm., T: ambient, U_G: 0.0014–0.14 m/s	$\varepsilon_G = 0.728U - 0.485U^2 + 0.0975U^3$ $$U = U_G\left[\dfrac{\rho_L^2}{\sigma_L(\rho_L-\rho_G)g}\right]^{\frac{1}{4}}$$
Reilly et al. (1986)	Air/H$_2$O, solvent, TCE/glass	P: atm., T: ambient, U_G: 0.02–0.2 m/s, C_V: up to 10 vol.%, d_c: 0.3 m, HC: 5 m	$\varepsilon_G = 296U_G^{0.44}\,\rho_L^{-0.98}\,\sigma_L^{-0.16}\,\rho_G^{0.19} + 0.009$
Sauer and Hempel (1987)	Air/H$_2$O/10 different solids $(1020<pp<2780$ kg/m$^3)$	P: atm., T: ambient, U_G: 0.01–0.08 m/s, C_V: 0–20 vol.%	$\dfrac{\varepsilon_G}{1-\varepsilon_G} = 0.0277\left[\dfrac{U_G}{(U_G g\nu_{sl})^{0.25}}\right]^{0.844}\left(\dfrac{\nu_{sl}}{\nu_{eff.rad}}\right)^{-0.136}\left(\dfrac{C_s}{C_{sp}}\right)^{0.0392}$ where C_{sb} is solid concentration at the bottom of column (kg/m^3) $$\nu_{sl} = \dfrac{\mu_L\left[1+2.5C_V+10.05C_V^2+0.00273\exp(16.6C_V)\right]}{\rho_{SL}}$$

TABLE 3.5 (Continued)

Author	System studied	Range of Operating Variables	Available correlation(s)
Schumpe et al. (1987)	N$_2$, O$_2$/H$_2$O, 0.8 M Na$_2$SO$_4$/carbon, kieselguhr, aluminium oxide	P: atm., T: ambient, U_G: up to 0.07 m/s, C_s: up to 300 kg/m³, d_C: 0.095 m, H_C: 0.85 m	$v_{eff,rad} = 0.011 D_C \sqrt{gD_C} \left(\dfrac{U_G^3}{gv_L} \right)^{\frac{1}{8}}$, $\varepsilon_G = BU_G^{0.87} \mu_{eff}^{-0.18}$, $\mu_{eff} = k(2800U_G)^{n-1}$; k and n are F ($C_V$, solid nature), B=0.81 or 0.43, 0.89≤103 K (Pa sn)≤1730, 0.163≤n≤1
Öztürk et al. (1987)	Air/water, ligroin, tetralin, aqueous solution Na$_2$SO$_4$ (0.8 M)/PE (24.6 and 106 μm), PVC (82 μm), activated carbon (5.4 μm), kieselguhr (6.6 μm), Al$_2$O$_3$ (10.5 μm)	P: 0.1 MPa; T: 293 K; U_G: 0–0.08 m/s; U_L: 0 m/s; C_V: 0%	$\varepsilon_G = K_1 \mu_G^{0.77} \mu_{SL}^{-0.21}$ K_1 depends on liquid-phase properties
Reilly et al. (1994)	Air Ar, He/water, Varsol, Varsol + antifoam, trichloroethylene/glass beads (71–745 μm)	P: 0.1 MPa; T: 283–323 K; U_G: 0–0.35 m/s; U_L: 0 m/s; C_V: 0–25%	$\varepsilon_G = 296 u_G^{0.44} \sigma_L^{-0.16} \rho_L^{-0.98} \rho_G^{0.19} + 0.009$
Fan et al. (1999)	N$_2$/Paratherm NF/alumina	P: 0.1–5.62 MPa, T: 301 and 351 K, U_G up to 0.45 m/s, C_V: 8.1,19.1 vol. %, d_C: 0.102 m, H_C: 1.37 m	$\dfrac{\varepsilon_G}{1-\varepsilon_G} = \dfrac{2.9 \left(\dfrac{U_G^4 \rho_G}{\sigma_L g} \right)^{\alpha} \left(\dfrac{\rho_G}{\rho_{SL}} \right)^{\beta}}{\left[\cosh \left(Mo_{SL}^{0.054} \right) \right]^{4.1}}$

TABLE 3.5　(Continued)

Author	System studied	Range of Operating Variables	Available correlation(s)
			$Mo_{SL} = g\left(\dfrac{(\rho_{SL}-\rho_G)(\xi\mu_L)^4}{\rho_{SL}^2\sigma_L^3}\right)$, $\alpha = 0.21 Mo_{SL}^{0.0079}$, and $\beta = 0.096 Mo_{SL}^{-0.011}$ $Ln\xi = 4.6C_v\left\{5.7C_v^{-0.58}\sinh\left[-0.71\exp(-5.8C_v)\ln Mo^{0.22}\right]+1\right\}$
Luo et al. (1999)	N_2/Paratherm NF/ alumina (100 μm)	P: 0.1–5.6 MPa; T: 301–351 K; U_G: 0–0.045 m/s; U_L: 0 m/s; C_v: 0–19.1%	$\dfrac{\varepsilon_G}{1-\varepsilon_G} = \dfrac{2.9\left(\dfrac{u_G^4\rho_G}{\sigma_{L}g}\right)^{0.21 Mo_m^{0.0079}}\left(\dfrac{\rho_G}{\rho_{SL}}\right)^{0.096 Mo_m^{-0.011}}}{\left[\cosh\left(Mo_m^{0.054}\right)\right]^{4.1}}$; $Mo_m = \dfrac{g(\rho_{SL}-\rho_G)\mu_{SL}^4}{\rho_{SL}^4\sigma_L^3}$
Sivaiah et al. (2012)	Air/water/zinc oxide	P: atmospheric; T: 295 K; U_g: 0–0.034 m/s; U_{sr}: 0–0.13 m/s; C_v: 0.1–1.0	$\varepsilon_g = 1-\exp\left(\dfrac{u_s-b}{a}\right)$; $a = 0.3901+0.0051w+0.0075w^2$ $b = 0.2403-0.003w-0.0005w^2$ $\varepsilon_{g,c} = 1-e^{(-b/a)} = $ critical gas hold-up at zero slip
Sivaiah and Majumder (2012)	Air/water/aluminium oxide, kieselgur	P: atmospheric; T: 295 K; U_g: 0–0.034 m/s; U_{sr}: 0–0.13 m/s; C_v: 0.1–1.0	$\varepsilon_G = \dfrac{1}{\left(1+\left(\dfrac{\rho_G}{\rho_{SL}}\right)\left(1.007\times10^{10}\,Re_{SLb}^{-0.249}\,Re_G^{-1.205}\,D_R^{-0.136}\,X^{-2.253}\right)\right)}$

TABLE 3.5 *(Continued)*

Author	System studied	Range of Operating Variables	Available correlation(s)
Sivaiah and Majumder (2013)	Air/water, electrolyte solution/zinc oxide, aluminium oxide, kieselgur	P: atmospheric; T: 295 K; U_g: 0–0.034 m/s; U_{sl}: 0–0.13 m/s; C_v: 0.1–1.0	$$\varepsilon_g = \frac{u_g}{C_o(u_g + u_{sl}) + V_d} ; \quad V_d = \sqrt{2}\left(\frac{g\sigma_l(\rho_l - \rho_g)}{\rho_l^2}\right)^{1/4} ;$$ $$C_o = 81.712 \mathrm{Re}_{sl}^{-1.366}\, \mathrm{Re}_g^{0.075}\, D_r^{-0.014}\, M_o^{-0.328}$$

KEYWORDS

- gas hold-up
- gas disengagement technique
- conductimetry
- downflow gas-interacting slurry reactor
- slip velocity model
- drift flux model
- Lockhart–Martinelli model
- homogeneous flow model
- variable-density model
- momentum exchange model
- Baroczy correlation
- Kawase and Moo-Young model

REFERENCES

Abdulrahman M. W. Experimental Studies of Gas Holdup in a Slurry Bubble Column at High Gas Temperature of a Helium–Water–Alumina System. *Chem. Eng. Res. Des.* **2016,** *109*(486), 494.

Akita, K.; Yoshida, F. Gas Holdup and Volumetric Mass Transfer Coefficient in Bubble Columns. *Ind. Eng. Chem. Process Des. Dev.* **1973,** *12,* 76–80.

Akita, K.; Yoshida, F. Bubble Size, Interfacial Area and Liquid-Phase Mass Transfer Coefficient in Bubble Columns. *Ind. Eng. Chem. Process Des. Dev.* **1974,** *13*, 84–91.

Bando, Y.; Kuraishi, M.; Nishimura, M.; Hattori, M.; Asada, T. Co Current Downflow Bubble Column with Simultaneous Gas-Liquid Injection Nozzle. *J. Chem. Eng. Jpn.* **1988,** *21*, 607–612.

Bando, Y.; Nishimura, M.; Sota, H.; Hattori, M.; Sakai, N.; Kuraishi, M. Flow Characteristics of Three Phase Fluidized Bed with Draft Tube-Effect of Outer Column Diameter and Determination of Gas-Liquid Interfacial Area. *Chem. Eng. J.* **1990,** *23*, 587–592.

Bankoff, S. G. A Variable Density Single-Fluid Model for Two-Phase Flow with Particular Reference to Steam—Water Flow. *J. Heat Transfer* **1960,** *82*(4), 265–272.

Baroczy, C. J. *Correlation of Liquid Fraction in Two-Phase Flow with Application to Liquid Metals*, NAA-SR-8171 1963.

Behkish, A.; Lemoine, R.; Oukaci, R.; Morsi, B. I. Novel Correlations for Gas Holdup in Large Scale Slurry Bubble Column Reactors Operating Under Elevated Pressures and Temperatures. *Chem. Eng. J.* **2006,** *115*, 157–171.

Behkish, A.; Lemoine, R.; Sehabiague, L.; Oukaci, R.; Morsi, B. I. Gas Holdup and Bubble Size Behavior in a Large-Scale Slurry Bubble Column Reactor Operating with an Organic Liquid Under Elevated Pressures and Temperatures. *Chem. Eng. J.* **2007**, *128*, 69–84.

Behkish, A.; Men, Z.; Inga, J. R.; Morsi, B. I. Mass Transfer Characteristics in a Large-Scale Slurry Bubble Column Reactor with Organic Liquid Mixtures. *Chem. Eng. Sci.* **2002**, *57*, 3307–3324.

Boyer, C.; Duquenne, A.; Wild, G.; August. Measuring Techniques in Gas—Liquid and Gas—Liquid Solid Reactors. *Chem. Eng. Sci.* **2002**, *57*(16), 3185–3215.

Briens, C. L.; Huynh, L. X.; Large, J. F.; Catros, A.; Bernard, J. R.; Bergougnou, M. A. Hydrodynamics and Gas—Liquid Mass Transfer in a Downward Venturi-Bubble Column Combination. *Chem. Eng. Sci.* **1992**, *47*, 3549–3556.

Butterworth, D. A Comparison of Some Void-Fraction Relationships for Co-Current Gas-Liquid Flow. *Int. J. Multiphase Flow* **1975**, *1*, 845–850.

Camarasa, E.; Vial, C.; Poncin, S.; Wild, G.; Midoux, N.; Bouillard, J. Influence of Coalescence Behavior of the Liquid and of Gas Sparging on Hydrodynamics and Bubble Characteristics in a Bubble Column. *Chem. Eng. Proc.* **1999**, *38*, 329–344.

Chen, Y. M.; Fan, L. S. Drift Flux in Gas-Liquid-Solid Fluidized Systems from the Dynamics of Bed Collapse. *Chem. Eng. Sci.* **1990**, *45*, 935–945.

Clark, N. N.; Flemmer, R. L. Predicting the Holdup in Two-Phase Bubble Upflow and Down Flow Using the Zuber and Findlay Drift Flux Model. *AIChE J.* **1985**, *31*, 500–503.

Darton, R. C.; Harrison, D. Gas and Liquid Hold-Up in Three-Phase Fluidization. *Chem. Eng. Sci.* **1975**, *30*, 581–586.

Deshpande, N. S.; Dinkar M.; Joshi, J. B. Disengagement of the Gas Phase in Bubble Columns. *Int. J. Multiphase Flow* **1995**, *21*(6), 1191–1201.

Douek, R. S.; Hewitt, G. F.; Livingston, A. G. Hydrodynamics of Vertical Co-Current Gas-Liquid-Solid Flows. *Chem. Eng. Sci.* **1997**, *52*, 4357–4372.

Douek, R. S.; Livingston, A. G.; Johansson, A. C.; Hewitt, G. F. Hydrodynamics of an External-Loop Three-Phase Airlift (TPAL) Reactor. *Chem. Eng. Sci.* **1994**, *49*, 3719–3737.

Evans, G. M.; Jameson, G. J.; Atkinson, B. W. Prediction of the Bubble Size Generated by a Plunging Liquid Jet Bubble Column. *Chem. Eng. Sci.* **1992**, *47*, 3265–3272.

Fan L-S; Yang, G. Q.; Lee, D. J.; Tsuchiya, K.; Luo, X. Some Aspects of High-Pressure Phenomena of Bubbles in Liquids and Liquid—Solid Suspensions. *Chem. Eng. Sci.* **1999**, *54*, 4681–4709.

Götz, M.; Lefebvre, J.; Mörs, F.; Reimert, R.; Graf, F.; Kolb, T. Hydrodynamics of Organic and Ionic Liquids in a Slurry Bubble Column Reactor Operated At Elevated Temperatures. *Chem. Eng. J.* **2016**, *286*, 348–360.

Hills, J. H. The Operation a of Bubble Column At High Throughputs. I. Gas Holdup Measurements. *Chem. Eng. J.* **1976**, *12*, 89–99.

Isbin, H. S.; Sher, N. C.; Eddy, K. C. Void Fractions in Two-Phase Steam—Water Flow. *AIChE J.* **1957**, *3*, 136–142.

Ishibashi, H.; Onozaki, M.; Kobayashi, M.; Hayashi J-i; Itoh, H.; Chiba, T. Gas Holdup in Slurry Bubble Column Reactors of a 150 T/D Coal Liquefaction Pilot Plant Process. *Fuel* **2001**, *80*, 655–664.

Jamialahmadi, M.; Muller-Steinhagen, H. Gas Hold-Up in Bubble Column Reactors. In *Encyclopedia of Fluid Mechanics;* Cheremisinoff, N. P., Ed.; Gulf Publishing Company: Houston, 1993, 387–407.

Jordan, U.; Schumpe, A. The Gas Density Effect on Mass Transfer in Bubble Columns with Organic Liquids. *Chem. Eng. Sci.* **2001**, *56*, 6267–6272.

Kastanek, F.; Zahradnik, J.; Kratochvil, J.; Cermak, J. Modeling of Large-Scale Bubble Column Reactors for Non-Ideal Gas-Liquid Systems. In *Frontier in Chemical Reaction Engineering;* Doraiswamy, L. K., Mashelkar, R. A., Eds.; John Wiley and Sons: New Delhi, India, 1984; Vol. 1, pp 330–344.

Kawanishi, K.; Hirao, Y.; Tsuge, A. An Experimental Study on Drift Flux Parameters for Two-Phase Flow in Vertical Round Tubes. *Nucl. Eng. Des.* **1990**, *120*, 447–458.

Kawase, Y.; Moo-Young, M. Theoretical Prediction of Gas Hold-Up in Bubble Columns with Newtonian and Non-Newtonian Fluids. *Ind. Eng. Chem. Res.* **1987**, *26*(5), 933.

Kemoun, A.; Cheng, B. O.; Gupta, P.; Al-Dahhan, M. H.; Dudukovic, M. P. Gas Holdup in Bubble Columns at Elevated Pressure via Computed Tomography. *Int. J. Multiphase Flow* **2001**, *27*, 929–946.

Koide, K.; Takazawa, A.; Komura, M.; Matsunaga, H. Gas Holdup and Volumetric Liquid-Phase Mass Transfer Coefficient in Solid-Suspended Bubble Columns. *J. Chem. Eng. Jpn.* **1984**, *17*, 459–466.

Kumar, A.; Degaleesan, T. E.; Laddha, G. S.; Hoelscher, H. E. Bubble Swarm Characteristics in Bubble Columns. *Can. J. Chem. Eng.* **1976**, *54*, 503–508.

Kumar, S. B.; Moslemian, D.; Duducovic, M. P. Gas-Holdup Measurements in Bubble Columns Using Computed Tomography. *AIChE J.* **1997**, *43*, 1414–1425.

Lapidus, L.; Elgin, J. C. Mechanics of Vertical-Moving Fluidized Systems. *AIChE J.* **1957**, *3*, 63–68.

Levy, S. Prediction of Two-Phase Annular Flow with Liquid Entrainment. *Intern. J. Heat Mass Transfer* **1966**, *9*, 171–188.

Lockett, M.; Kirkpatrick, R. D. Ideal Bubbly Flow and Actual Flow in Bubble Columns. *Trans. Inst. Chem. Eng.* **1975**, *53*, 267–273.

Lockhart, R. W.; Martinelli, R. C. Proposed Correlation of Data for Isothermal Two-Phase, Two Component Flow in Pipes. *Chem. Eng. Prog.* **1949**, *45*, 39–48.

Luo, X.; Lee, D. J.; Lau, R.; Yang, G. Q.; Fan, L. S. Maximum Stable Bubble Size and Gas Holdup in High-Pressure Slurry Bubble Columns. *AIChE J.* **1999**, *45*, 665–680.

Majumder, S. K., *Hydrodynamics and Transport Processes of Inverse Bubbly Flow,* 1st ed.; Elsevier: Amsterdam, 2016; p 192.

Majumder, S. K.; Kundu, G.; Mukharjee, D. Efficient Dispersion in a Modified Two-Phase Non Newtonian Downflow Bubble Column. *Chem. Eng. Sci.* **2006**, *61*, 6753–6764.

Marchese, M. M.; Uribe-Salas, A.; Finch, J. A. Measurement of Gas Holdup in a Three-Phase Concurrent Inverse Flow Column. *Chem. Eng. Sci.* **1992**, *47*, 3475–4382.

Marrucci, G. Rising Velocity of a Swarm of Bubbles. *Ind. Eng. Chem. Fun.* **1965**, *4*, 224–229.

Mowla, A.; Treeratanaphitak, T.; Budman, H. M.; Abukhdeir, N. M.; Ioannidis, M. A. A Meta-Analysis of Empirical Correlations for Average Gas Hold-Up in Three-Phase Fluidized Beds. *Powder Technol.* **2016**, *301*, 590–595.

Nassos, G.; Bankoff, S. G. Slip Velocity Ratios in an Air-Water Systems Under Steady State and Transient Conditions. *Chem. Eng. Sci.* **1967**, *22*, 681–688.

Ohkawa, A.; Kusabiraki, D.; Kawai, Y.; Sakai, N. Some Flow Characteristics of a Vertical Liquid Jet System Having Downcomers. *Chem. Eng. Sci.* **1986**, *41*, 2347–2361.

Öztürk, S. S.; Schumpe, A.; Deckwer, W. D. Organic Liquids in a Bubble Column: Holdups and Mass Transfer Coefficients. *AIChE J.* **1987**, *33*(9), 1473–1480.

Prandtl, L. Bemerkungen Zur Theorie Der Freien Turbulenz. *ZAMM* **1942**, *22*, 241–243, (cited by Sato et al., 1981).

Reilly, I. G.; Scott, D. S.; De Bruijn, T. J. W.; MacIntyre, D. The Role of Gas Phase Momentum in Determining Gas Holdup and Hydrodynamic Flow Regimes in Bubble Column Operations. *Can. J. Chem. Eng.* **1994**, *72*(1), 3–12.

Reilly, J. G.; Scott, D. S.; Bruijn, T.; Jain, A.; Diskorz, J. Correlation for Gas Holdup in Turbulent Coalescing Bubble Columns. *Can. J. Chem. Eng.* **1986**, *64*, 705–717.

Sarrafi, A.; Jamialahmadi, M.; Muller-Steinhagen, H.; Smith, J. M. Gas Holdup in Homogeneous and Heterogeneous Gas-Liquid Bubble Column Reactors. *Can. J. Chem. Eng.* **1999**, *77*(1), 11–21.

Sauer, T.; Hempel, D. C. Fluid Dynamics and Mass Transfer in a Bubble Column with Suspended Particles. *Chem. Eng. Technol.* **1987**, *10*, 180–189.

Schügerl, K.; Lücke, J.; Oels, U. Bubble Column Bioreactions. *Adv. Biochem. Eng.* **1977**, *7*, 1–84.

Schumpe, A.; Saxena, A. K.; Fang, L. K. Gas—Liquid Mass Transfer in a Slurry Bubble Column. *Chem. Eng. Sci.* **1987**, *42*, 1787–1796.

Shah, M.; Kiss, A. A.; Zondervan, E.; Van der Schaaf, J.; De Haan, A. B. Gas Holdup, Axial Dispersion, and Mass Transfer Studies in Bubble Columns. *Ind. Eng. Chem. Res.* **2012**, *51*, 14268–14278.

Shah, Y. T.; Kulkarni, A. A.; Wieland, J. H. Gas Holdup in Two and Three-Phase Downflow Bubble Columns. *Chem. Eng. J.* **1983**, *26*, 95–104.

Shen, G.; Finch, J. A. Bubble Swarm Velocity in a Column. *Chem. Eng. Sci.* **1996**, *51*, 3665–3674.

Sivaiah, M.; Parmar, R.; Majumder, S. K. Gas Entrainment and Holdup Characteristics in a Modified Gas-Liquid-Solid Down Flow Three-Phase Contactor. *Powder Technol.* **2012**, *217*, 451–461.

Sivaiah, M.; Majumder, S. K. Hydrodynamics and Mixing Characteristics in an Ejector-Induced Downflow Slurry Bubble Column (Eidsbc). *Chem. Eng. J.* **2013**, *225*, 720–733.

Sivaiah, M.; Majumder, S. K. Gas Holdup and Frictional Pressure Drop of Gas-Liquid-Solid Flow in a Modified Slurry Bubble Column. *Int. J. Chem. Reactor Eng.* **2012**, *10*, 1–29.

Sriram, K.; Mann, R. Dynamic Gas Disengagement: A New Technique for Assessing the Behaviour of Bubble Columns. *Chem. Eng. Sci.* **1977**, *32*, 571–580.

Ueyama, K.; Miyauchi, T. Properties and Recirculating Turbulent Two Phase Flow in Gas Bubble Columns. *AIChE J.* **1979**, *25*, 258–266.

Uribe-Salas, A.; Gomez, C. O.; Finch, J. A. A. Conductivity Technique for Gas and Solids Holdup Determination in Three-Phase Reactors. *Chem. Eng. Sci.* **1994**, *49*, 1–10.

Vail, Y. K.; Manokov, N. Kh.; Manshilin, V. V. The Gas Contents of Three-Phase Fluidized Beds. *Int. Chem. Eng.* **1970**, *10*, 244.

Wachi, S.; Yates, J. G. Downward Lean Powder Flow in a Vertical Tube. *Chem. Eng. Sci.* **1991**, *46*(929) 937.

Wilkinson, P. M.; Spek, A. P.; Van Dierendonck, L. L. Design Parameters Estimation for Scaleup of High-Pressure Bubble Columns. *AIChE J.* **1992**, *38*(4), 54–544.

Yamagiwa, K.; Kusabiraki, D.; Ohkawa, A. Gas Holdup and Gas Entrainment Rate in Downflow Bubble Column with Gas Entrainment by a Liquid Jet Operating At High Liquid Throughput. *J. Chem. Eng. Jpn.* **1990**, *23*, 343–348.

Zahradnik, J.; Fialova, M.; Linek, V.; Sinkule, J.; Reznickova, J.; Kastanek, F. Dispersion Efficiency of Ejector Type Gas Distributor in Different Operating Modes. *Chem. Eng. Sci.* **1997,** *52,* 4499–4510.

Zhang, K.; Qi, N.; Jin, J.; Lu, C.; Zhang, H. Gas Holdup and Bubble Dynamics in a Three-Phase Internal Loop Reactor with External Slurry Circulation. *Fuel* **2010,** *89,* 1361–1369.

Ziganshin, G. K.; Ermakova, A. Gas Content with the Ascending Direct Flow of a Gas-Liquid Stream in the Presence of Fixed and Fluidized Beds of Granular Material. *Theor. Found. Chern. Eng.* **1970,** *4,* 576.

Zuber, N.; Findlay, J. A. Average Volume Concentration in Two-Phase Flow Systems. *J. Heat Trans.* **1965,** *87,* 453–468.

CHAPTER 4

FRICTIONAL PRESSURE DROP

CONTENTS

4.1   INTRODUCTION

The downflow gas-interacting slurry bubble column is gaining importance due to its several advantages as discussed in Chapter 1. It is not yet recommended for the industrial applications due to the lack of understanding of complex multiphase phenomena such as hold-up, pressure drop, mixing and heat transfer in the downflow system. Pressure drop is an important design parameter in the three-phase flow. In the slurry bubble column reactor, the reaction between gas and liquid in the presence of catalyst particle occurs under certain pressure and temperature. The influence of pressure seems to be higher on the heterogeneous reactive system than in the homogeneous reactive system. In the gas slurry reaction, the performance depends on how the interfacial area is produced and the volume fraction of the gas bubble. The interfacial area depends on the bubble size. The interfacial area is increased with decrease in the bubble size. The bubble diameter decreases when the pressure increases. Pressure increase results in an increase of gas volume fraction. It is also important to say that gas–liquid distribution is crucial for the effective gas–slurry dispersion and contacting, thereby ensuring a good reactor performance and facilitating reactor scale-up. The volume fraction of the gas increases with superficial velocity of the gas in the slurry bubble column reactor due to the decrease in the hydrostatic pressure with increasing gas volume fraction in the reactor. With higher values of gas flow rate, the bed pressure drop decreases further because of the reduction in the hydrostatic pressure (Abdulrahman, 2016). For gas–liquid-solid system, the observed bed pressure drop recorded in pressure transducers was reported to be less than that for liquid–slurry bed only. Also, for a given gas velocity, the hydrostatic pressure and the pressure drop increases by increasing height of the reactor, which leads to the decrease of gas hold-up. Furthermore, for the constant pressure and gas velocity, the gas momentum per unit mass of slurry decreases with solid concentration, and consequently, the overall gas hold-up is expected to decrease and leads to the decrease of slurry mixing rate and affect the transport processes in the slurry reactor. The gas hold-up does not change with tube diameter, solid concentration and average size of the particles except at low gas and higher superficial slurry velocity, whereas the frictional pressure drops increase with the concentration of the solid particles, but independent of average size of the particles reported by Hatate et al. (1986) and Sivaiah and Majumder (2013).

This positive effect of pressure on the gas hold-up is generally observed in heterogeneous reactive system (Pjontek et al., 2014). However, in the homogeneous system, an appreciable increase is reported in the literature (Shaikh and Al-Dahhan, 2005). The overall mass transfer coefficient increases with increasing pressure (Majumder, 2016). Jordan and Schumpe (2001) reported that the mass transfer increases with the gas density by increasing pressure. Wilkinson et al. (1994) explained that in homogeneous system, pressure would have a negligible effect such as the breakage and coalescence of the bubble, and hence, on the mass transfer characteristics. However, Maalej et al. (2003) reported a decrease of the overall volumetric mass transfer coefficient with pressure at a constant mass flow rate. A pressure increase results in an increase of the interfacial area (Han and Al-Dahhan, 2007). Han and Al-Dahhan (2007) observed the decrease of individual mass transfer coefficient of liquid (k_l) while increasing the pressure. The mass transfer coefficient (k_l) decreases by 20% between 0.1 and 0.4 MPa but remains constant between 0.4 and 1 MPa. The authors explained this trend with the Higbie's theory and reported that the increase in the pressure reduces the size of the bubbles. The authors argued that the small bubbles have a slower slip velocity, so the contact time would decrease. This would mean that the k_l decreases when the bubble size decreases (Leonard et al., 2015). This is contradictory with Kulkarni (2007) and his interpretation of the theory of Calderbank, which affirms exactly the opposite. In fact, if the slip velocity decreases when the diameter decreases, it is difficult to assess the evolution of the ratio d_b/u_s when both the terms decrease. Yang et al. (2001) reported no influence of pressure on the mass transfer coefficient k_l, at different temperatures (from 293 to 473 K) in the case of a quite viscous liquid paraffin (μ_l=0.26–3 Pa·s). The authors explained that the gas solubility increases with pressure. This increasing solubility results in more gas dissolved in the liquid, which leads to a decrease in viscosity. They, therefore, concluded that the k_l decreases when the bubble size decreases, but in parallel, it is favoured by the decrease in viscosity reduction that would increase the molecular diffusivity. Ultimately, the two effects cancel each other out and resulted constant is k_l (Leonard et al., 2015).

A decrease of the liquid axial dispersion coefficient with an increase of pressure in a wide range of superficial gas and liquid velocity is reported by Yang and Fan (2003) and Leonard et al. (2015). The effect is higher at high superficial gas velocity and the effect of pressure decreases when increasing pressure. The authors explained it by using fluid circulation model proposed by Joshi (1980), where the turbulence associated with liquid circulation results in the dispersion in case of the heterogeneous regime. However, the effect of pressure would be linked to several contradictory effects on fluid circulation velocity (Therning and Rasmuson, 2001; Ruthiya et al., 2005 and Chilekar et al., 2010; Leonard et al., 2015). In fact, liquid recirculation increases when increasing pressure. Houzelot et al. (1983) observed no effect of pressure on the liquid axial dispersion coefficient for pressures between 0.1 and 0.3 MPa in the homogeneous regime. Holcombe et al. (1983) also reported no effect of pressure in continuous mode at higher superficial gas velocities and for a limited range of pressure (0.3–0.71 MPa). The theory disagrees with Wilkinson et al. (1993) who observed an increase of the liquid axial dispersion coefficient with pressure between 0.1 and 1.5 MPa for the same nitrogen/water system but with a higher column diameter, though they observed an increase of the liquid axial dispersion coefficient with pressure despite the decrease of the bubble size (Leonard et al., 2015).

The liquid phase properties have an immense effect on the pressure and liquid axial dispersion coefficient. The energy distribution in the reactor may also affect the pressure drop and the generation of liquid circulation cell in the reactor due to bubble-induced turbulence. The liquid axial velocity decreases with increasing pressure in the central region of slurry bubble column. This occurs because the bubble size and bubbles' rising velocity decrease with an increase in the pressure (Luo et al., 1999) and results in a degree of bubble-induced turbulence. Therefore, the liquid axial velocity decreases with the increase of the operating pressure. The pressure effect is significant and the stress decreases as the pressure increases. The mean bubble size decreases with increasing pressure. The bubble size plays an important role in determining the fluctuations of the liquid phase. The fluctuations of the liquid phase induced by bubbles are reduced due to the size distribution of bubble. This affects the instantaneous reaction in slurry bubble reactor.

The gas-interacting three-phase pressure drop is an essential element for the design of process which is the subject of numerous experimental and numerical studies. To design and control such equipment efficiently, the values of pressure drops and volumetric fractions of phases must be determined. The influences of the temperature and pressure must be anticipated since all slurry reactors for gas-to-liquid processes operate at high pressure and temperature (Sivaiah and Majumder, 2013).

4.2   THEORY OF THREE-PHASE FRICTIONAL PRESSURE DROP

The total pressure drop comprises the frictional, gravitational and the pressure drop due to acceleration, which is represented by Equation 4.1

$$\Delta P_{tp} = \Delta P_{ftp} + \Delta P_{h} + \Delta P_{a} \tag{4.1}$$

where ΔP_{h} is the pressure drop due to the change in the potential energy. In the present gas-interacting jet-induced downflow slurry reactor, the hydrostatic pressure drop can be calculated once the fractional gas hold-up is known by the experiment by Marchese et al. (1992).

$$\Delta P_{h} = h_{m}\rho_{sl}(1-\varepsilon_{g})g \tag{4.2}$$

In the case of gas-interacting jet-induced downflow slurry reactor since the liquid jet impinges on the pool surface, the liquid aches a sudden braking, which enhances to raise the dynamic pressure in the reactor. The dynamic pressure (ΔP_{a}) is equal to the ratio of force exerted by the jet and the cross-sectional area of the column, which can be expressed as

$$\Delta P_{a} = \frac{Q_{sl}\rho_{sl}\left(u_{j} - u_{sl}\right)}{A_{c}} \tag{4.3}$$

where u_{sl} and u_{j} are the interstitial slurry and jet velocities, respectively, and Q_{sl} is the volumetric slurry flow rate. Equation 4.3 can be further simplified as

$$\Delta P_{a} = u_{sl}^{2}\rho_{sl}\left[\frac{d_{c}^{2}}{d_{j}^{2}} - \frac{1}{\left(1-\varepsilon_{g}\right)}\right] \tag{4.4}$$

From Equations 4.1–4.3, the three-phase frictional pressure drop (ΔP_{ftp}) can be represented as

$$\Delta P_{ftp} = \Delta P_{tp} - h_{m}\rho_{sl}\left(1-\varepsilon_{g}\right)g - \left(u_{sl}^{2}\rho_{sl}\left[\frac{d_{c}^{2}}{d_{j}^{2}} - \frac{1}{\left(1-\varepsilon_{g}\right)}\right]\right) \tag{4.5}$$

The frictional pressure drop can also be expressed based on the actual slurry velocity as

$$\frac{\Delta P_{ftp}}{g\rho_{sl}h_m} = \frac{2f_{tpl}u_{sl}^2}{gd_c}$$ (4.6)

By substituting Equation 4.6 into Equation 4.5, the three-phase friction factor (f_{tpl}) can be obtained as

$$f_{tpl} = \left(\frac{gd_c}{2u_{sl}^2}\right)\left(\frac{\Delta P_{tp}}{g\rho_{sl}h_m} - (1-\varepsilon_g) - \left(\frac{u_{sl}^2}{gh_m}\left[\frac{d_c^2}{d_j^2} - \frac{1}{(1-\varepsilon_g)}\right]\right)\right)$$ (4.7)

In the vertical flow, it is usual that the gravitational component constitutes a major part of the total pressure drop. However, the gravitational component can be assumed negligible in high gas volume fraction flows (Sani, 1960). Thus, an accurate measurement of the gravitational component is a prerequisite for the determination of the frictional pressure drop in such systems (Kiausner et al., 1990). Kiausner et al. (1990) reported that the two-phase frictional pressure drop for the downflow behaves in a characteristically different manner than that of the upflow, and it is significantly higher in the adiabatic system than in the nonadiabatic system at a certain gas hold-up.

4.3   EFFECTS OF OPERATING VARIABLES ON FRICTIONAL PRESSURE DROP

The frictional pressure drop increased with the increase in slurry Reynolds number and decreased with the increase in gas Reynolds number as shown in Figure 4.1 (Sivaiah and Majumder, 2013). Majumder (2013) reported that with increasing gas flow rate, the flow area of the slurry inside the column decreases and also the coalescence of the bubbles inside the column lead to generate larger bubbles. This leads to decrease the friction in the column. They also reported that the variation of frictional pressure drop is higher at higher slurry flow rates as compared to the lower slurry flow rates. The finer particles cause more friction than the coarser particles as they try to adhere to the walls and have higher tendencies of cluster formations (Sivaiah and Majumder, 2013). Also, the friction due to the finer particles is more at higher slurry flow rates because of the increase in turbulence of the gas–slurry–solid mixing and decrease in the area of the gas flow inside the column.

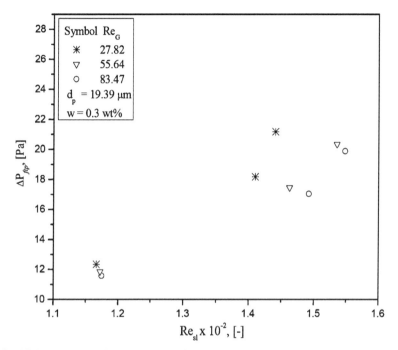

FIGURE 4.1   Frictional pressure drop with slurry and gas Reynolds number.

The variation of frictional pressure drop with slurry concentration at constant particle diameter (19.39 μm) is shown in Figure 4.2. It is observed that the frictional pressure drop decreases with an increase in the slurry concentration. With an increase in the concentration of the slurry, the apparent viscosity of the slurry increases which decreases the gas hold-up in the column. The decrease in gas hold-up increases the hydrostatic pressure and consequents the decrease in frictional pressure drop in the column. Also, the intensity of the mixing of gas–slurry–solid mixture and the movement of the liquid and particles are hindered, which results in the decrease of the friction in the column. With increasing the slurry concentration, the particle fluctuations and velocity would be reduced due to the higher friction losses between the particle and the viscous liquid medium. The frictional pressure drop in the downflow slurry bubble column was more for the finer particles as compared to the bigger particles as shown in Figure 4.3.

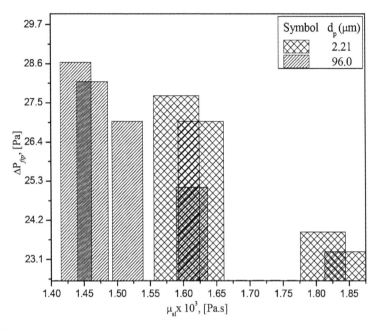

FIGURE 4.2   Variation of frictional pressure drop with slurry viscosity.

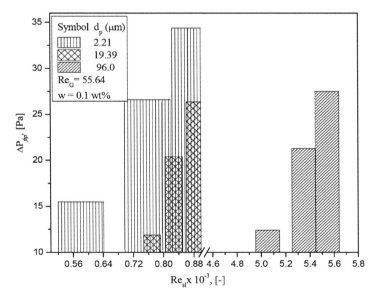

FIGURE 4.3   Variation of frictional pressure drop with slurry Reynolds number at different particle diameter.

At a constant gas flow rate, with an increase in the slurry flow rate, the intensity of mixing leads to increase the particle mobility and velocity inside the column. This particle mobility is more for finer particles as compared to bigger size particles, which leads to an increase in the frictional pressure drop with slurry flow rate. Xu et al. (2009), as per the result of Roy and Al-Dahhan (2005), reported that the liquid distribution in a slurry bed by comparing three different types of distributors (nozzle, showerhead and foam); the nozzle-type distributor is most suitable for the monolith three-phase reactors.

Xu et al. (2009) explained the significance of the choice of a distributor for uniform distribution of the phase. The distributor affects the pressure energy distribution in the slurry reactor based on the distribution of the phase hold-up in the reactor. They correlated the multiphase hydrodynamics of the monolith packing and the configuration of distributors in the three-phase reactor. They developed correlations for estimating the friction pressure drop and liquid hold-up, and an analysis of the unstable flow phenomena characterized by negative pressure drops over the packing. Ong and Bhatia (2009) studied the suitability of composite catalyst containing H–Y zeolite and kaolin for catalytic cracking of palm oil for the production of biofuels synthesized by sol–gel–sieve method. They reported that the pressure difference across the transport riser reactor changes with the superficial gas velocity, and it becomes significant at the curve section of the reactor. They suggested a model by tying different dimensionless groups correlated with the pressure drop in the reactor with different operating variables. The pressure drop characteristics play an important role in the design of transport slurry reactor. Kunii and Levenspiel (1991) reported that the reactor diameter influences particle size on the total pressure drop across the whole riser and the differential pressure gradient with apparent solids concentration at a specified bed height in the catalytic slurry reactor. With increasing riser diameter, the total pressure drop across the whole riser and differential pressure gradient increased with Geldart B-type particle but decreased for Geldart A-type particles (Ong and Bhatia, 2009; Xu et al., 1999). Besides, the variation in mean particle diameter and superficial velocity do affect the pressure, especially in the core region. Radial pressure profile gets flatter in the core region as the mean particle diameter increases. Gungor and Eskin (2006) reported the impact on the total pressure drop by the friction of gas and solids, which is insignificantly small compared to the acceleration and solid's hydrodynamic head components. Satterfield and Ozel (1977) reported that the pressure drop is profound to the design of distributor according to their conclusion of studies with three different distributors. Bukur et al. (1996) elucidated that the coalescence rate is close to the breakup rate pointing to a fully developed flow characterized by insignificant gas hold-up axial gradients in the slurry column at high-pressure conditions. In fact, at high pressure, due to the Kelvin–Helmholtz instability, the bubbles diameter in the gas–liquid–solid slurry reactor decreases to conserve their integrity (Krishna, 2000). Iliuta et al. (2008) analysed the pressure gradient in a slurry reactor by multicomponent hydrodynamic model based on the experimental works of other investigators. They reported that the pressure drop is very delicate to the catalyst concentration, exclusively at low gas density. This may be a reason for the increase of pressure drop with catalyst concentration. They also reported that significant reduction in large bubbles diameter with pressure implies an important reduction of their adjacent velocity. Transition velocity also has a significant effect on the pressure. Most of the researchers have described that transition velocities for flow regime in the slurry reactor increase with increasing pressure (Reilly et al., 1994; Shaikh and M. Al-Dahhan, 2005; Lin et al., 2001; Chilekar et al., 2010). Krishna et al. (1991) have performed extensive sets of experiments at pressures ranging from 0.1 to 2 MPa and with several gases (nitrogen, carbon dioxide, argon, helium and sulphur hexafluoride) in deionized water with 0.16 m diameter and 1.2 m height bubble column. They reported that the superficial gas velocity at the regime transition point was found to be a unique function of the gas density for all experiments in water, covering both effects of pressure and molar mass. Understanding the flow regimes in three-phase bubble column systems is very important for reactor design and scale-up (Nedeltchev et al., 2006). Different flow regimes affect reactor performance in many respects such as pressure fluctuation, mass transfer, heat transfer, momentum loss, mixing and reactor volume productivity (Nedeltchev et al., 2006). Barghi et al. (2004) used pressure fluctuation analysis to identify flow regimes in a slurry bubble column. They observed a free-bubbling regime when the gas velocity was below 0.05 m/s and gross recirculation patterns when the gas velocity was above 0.125 m/s. The methods for pressure fluctuation analysis under time domain analysis reveal the linear characteristics of a system. Ruthiya et al. (2005) developed an unambiguous flow regime transition identification method based on the coherent standard deviation and the average frequency analysis of pressure fluctuations in the slurry bubble columns. Li et al. (2013) investigated the flow regime transitions in a gas–liquid–solid three-phase bubble column based on pressure time series. They applied the statistical, Hurst, Hilbert–Huang transform and Shannon entropy analysis methods to differential pressure fluctuation data measured in a two-dimensional (2D) bubble column measuring 0.1 m in length and 0.01 m in width equipped with a sintered plate distributor (average diameter of holes was 50 µm). They successfully identified two-flow regime transition gas velocities based on sudden changes in both the empirical mode decompositions (EMD) energy entropy from Hilbert–Huang transform and the Shannon entropy values. They reported that the homogeneous regime shifted to the transition regime at a superficial gas velocity of 0.069 m/s and the transition regime shifted to the heterogeneous regime at a superficial gas velocity of 0.156–0.178 m/s. They concluded that the EMD energy entropy and Shannon entropy analysis methods can reveal the complex hydrodynamics underlying gas–liquid–solid flow and are confirmed to be reliable and efficient as non-invasive methods for detecting the flow regime transitions in the three-phase bubble column systems.

4.4 ANALYSIS OF FRICTIONAL PRESSURE DROP

4.4.1 LOCKHART–MARTINELLI MODEL

One of the most widely used recognized tools to describe the three-phase pressure drop in different process equipment is Lockhart–Martinelli correlation (Lockhart and Martinelli, 1949; Chisholm, 1967; Azbel and Kemp-Pritchard, 2009). Originally, Lockhart and Martinelli (1949) had proposed a graphical correlation for the analysis of frictional pressure drop in horizontal two-phase gas–slurry system for four different combinations, that is, laminar–laminar, laminar–turbulent, turbulent–laminar and turbulent–turbulent of gas and liquid flow. They defined the parameters φ_{sl}, φ_g and X as follows:

$$\Delta P_{ftp} = \phi_{sl}^2 \Delta P_{fosl} \tag{4.8}$$

$$\Delta P_{ftp} = \phi_g^2 \Delta P_{fog} \tag{4.9}$$

and

$$X = \frac{\phi_g}{\phi_{sl}} = \sqrt{\left(\frac{\Delta P_{fosl}}{\Delta P_{fog}}\right)} \tag{4.10}$$

The sole-phase frictional pressure drop ΔP_{fo} over a distance of Δz of the column can be evaluated by the relation

$$\Delta P_{fo} = \frac{2\rho_o f_o u_o^2 \Delta z}{d_c} \tag{4.11}$$

The single-phase friction factor (f_o) is calculated for laminar and turbulent flow in a hydraulically smooth tube (McCabe and Smith, 1976; Chhabra and Richardson, 1999), respectively, as

$$f_o = 16 / Re_o \tag{4.12}$$

and

$$f_o = 0.079 / Re_o^{0.25} \tag{4.13}$$

For transition flow, the single-phase friction factor (f_o) can be calculated as (Desouky and El-Emam, 1989)

$$f_o = 0.125 \left(0.0112 + Re_o^{-0.3185} \right) \tag{4.14}$$

Davis (1963) extended the applicability of Lockhart–Martinelli's correlation to the vertical flow through modification of the parameter X by incorporating the Froude number (Fr) to account for the effect of gravity and velocity. The modified parameter X_{mod} is expressed as

$$X_{mod} = 0.19 Fr^{0.185} X \tag{4.15}$$

where the Froude number is $Fr = u_m^2 / (gd_c)$. Chisholm (1967) developed the equations in terms of the Lockhart–Martinelli correlating groups for the friction pressure gradient during the flow of gas–slurry or vapour–liquid mixtures in the pipes. He developed a simplified equation for the use in engineering design as

$$\phi_{sl}^2 = f(X) = \left(1 + \frac{C}{X} + \frac{1}{X^2} \right) \tag{4.16}$$

$$\phi_g^2 = f(X) = \left(1 + CX + X^2 \right) \tag{4.17}$$

This parameter X is a function of the ratios of mass fluxes, densities and viscosities of the liquid and the gas phase in addition to the diameter of the pipe.

$$X^2 = \lambda \left(\frac{\dot{m}_l}{\dot{m}_g}\right)^a \left(\frac{\rho_g}{\rho_l}\right)^b \left(\frac{\mu_l}{\mu_g}\right)^c \qquad (4.18)$$

The constant λ and the coefficients a, b, c can be obtained by regression analysis of experimental data. The parameter C depends on the experimental conditions whether the liquid and gas phases are laminar or turbulent flow. The values of C are restricted to mixtures with gas–slurry density ratios corresponding to gas–slurry mixtures at atmospheric pressure. For turbulent and laminar flow of respective liquid and gas, the value of C is 10. Based on the Lockhart–Martinelli concept, for downflow gas-interacting slurry reactor (DGISR), the parameters, φ_{sl} and φ_g, have been correlated by incorporating the effect of different dimensionless groups developed by dimensional analysis as (Sivaiah and Majumder, 2012)

$$\varphi_{sl}^2 = 2.56 \times 10^2 \, Re_{slb}^{0.093} \, Re_g^{-0.551} \, D_r^{0.102} X^{-0.969} \qquad (4.19)$$

The correlation coefficient (R^2) and the standard error of Equation 4.19 were found to be 0.989 and 0.073, respectively. The predicted values of three-phase frictional pressure drop can then be calculated as:

$$\Delta P_{ftp} = \varphi_{sl}^2 \Delta P_{fosl} \qquad (4.20)$$

The frictional pressure drop multiplier (X) gives the satisfactory prediction for frictional pressure drop in the three-phase flow of the DGISR for the range of the variables of $22.27 < Re_{slb} < 575.2$, $27.82 < Re_g < 83.47$, $0.52 \times 10^3 < Dr < 22.68 \times 10^3$ and $16.13 < X < 64.99.2$.

4.4.2 OTHER CORRELATION MODELS

4.4.2.1 BANKOFF MODEL

The model is proposed for a pressure drop based on the phenomenology of interfacial shear stress of bubbly flow (Bankoff, 1960). The model equation for the frictional pressure gradient in terms of friction multiplier φ_{Bf} can be represented as

$$\left(\frac{dp}{dz}\right)_{f,tp} = \varphi_{Bf}^{7/4} \left(\frac{dp}{dz}\right)_{sl} \qquad (4.21)$$

The Bankoff multiplier in case of three-phase flow can be expressed as

$$\phi_{Bf} = \frac{1}{1-x}\left[1-\gamma\left(1-\frac{\rho_g}{\rho_{sl}}\right)\right]^{3/7}\left[1+x\left(\frac{\rho_{sl}}{\rho_g}-1\right)\right] \qquad (4.22)$$

where

$$\gamma = \left[c_1 + c_2\left(\frac{\rho_g}{\rho_{sl}}\right)\right]\left[1+\left(\frac{1-x}{x}\right)\left(\frac{\rho_g}{\rho_{sl}}\right)\right]^{-1} \qquad (4.23)$$

The coefficients c_1 and c_2 depend on the experimental data.

4.4.2.2 BAROCZY MODEL

Baroczy (1966) introduced property index (I_p) in terms of viscosity and density of phases, which can be represented for slurry–gas flow as

$$I_p = \left(\frac{\mu_{sl}}{\mu_g}\right)^{0.2} / \left(\frac{\rho_{sl}}{\rho_g}\right) \qquad (4.24)$$

The model interprets the phenomena when each phase flows alone turbulently. The inverse of the property index represents the ratio of the pressure drop gradient for all gas flow to that for all liquid flow which is given by

$$\left(\frac{\Delta p_f}{\Delta z}\right)_g / \left(\frac{\Delta p_f}{\Delta z}\right)_{sl} = \frac{1}{I_p} \qquad (4.25)$$

The model can also be used to analyse the pressure drop in slurry bubble column, provided all the physical properties of the system of slurry and gas are known.

4.4.2.3 WALLIS MODEL

Wallis (1969) proposed a correlation and reported good prediction results for bubbly flow, which is expressed in terms of the liquid friction factor multiplier. The value of liquid friction factor multiplier decreases with increasing system pressure at a given value of the flow quality. The proposed correlation can be used for three-phase system if the liquid is considered as slurry. The correlation can be represented in case of slurry–gas system as

$$\phi_W^2 = \left(1 + x\frac{\rho_{sl} - \rho_g}{\rho_g}\right)\left(1 + x\frac{\mu_{sl} - \mu_g}{\mu_g}\right)^{-0.25} \tag{4.26}$$

However, the correlation predicts the frictional pressure gradient well for the annular flow pattern. The correlation can be modified and applied in slurry–bubbly flow system by incorporating experimental data.

4.4.2.4   CHAWLA MODEL

The Chawla (1972) model can be used based on the gas pressure gradient in the slurry bubble column reactor:

$$\left(\frac{dp}{dz}\right)_{f,tp} = \varphi_{Chawla}\left(\frac{dp}{dz}\right)_g \tag{4.27}$$

The vapour-phase frictional gradient is calculated as

$$\left(\frac{dp}{dz}\right)_g = \frac{2f_g\dot{m}_t^2}{d_c\rho_g} \tag{4.28}$$

The Chawla two-phase multiplier depends on the slip ratio (S) and the multiplier, which can expressed in case of three-phase flow as

$$\varphi_{Chawla} = x^m\left[1 + S\left(\frac{1-x}{x}\right)\frac{\rho_g}{\rho_l}\right]^n \tag{4.29}$$

where the slip ratio can be expressed by developing correlations for the three-phase flow as

$$S = \frac{1}{\lambda\left[\dfrac{1-x}{x}(Re_g\,Fr_m)^{n_1}\left(\rho_{sl}/\rho_g\right)^{n_2}\left(\mu_{sl}/\mu_g\right)^{n_3}\right]} \tag{4.30}$$

where $Fr_m = \dot{m}_t^2/(gd_c\rho_m^2)$ and $Re_g = \dot{m}_g d_c/\mu_g$.

4.4.2.5   FRIEDEL MODEL

The Friedel correlation (Friedel, 1980) can be applied to three-phase flow in slurry reactor. The correlation can be expressed as

$$\left(\frac{dp}{dz}\right)_f = \phi_{Frd}\left(\frac{dp}{dz}\right)_{sl} \tag{4.31}$$

where

$$\phi_{Frd}^2 = E + \frac{\lambda FG}{Fr_m^{0.045}We_{sl}^{0.035}} \tag{4.32}$$

where

$$E = (1-x)^2 + x^2\frac{\rho_{sl}}{\rho_g}\frac{f_g}{f_{sl}} \tag{4.33}$$

$$F = (1-x)^{0.224}x^{0.78} \tag{4.34}$$

$$G = \left(\frac{\rho_{sl}}{\rho_g}\right)^{0.91}\left(\frac{\mu_g}{\mu_{sl}}\right)^{0.19}\left(1-\frac{\mu_g}{\mu_{sl}}\right)^{0.70} \tag{4.35}$$

$$Fr_m = \frac{\dot{m}_t^2}{gd_c\rho_m^2} \tag{4.36}$$

$$We_{sl} = \frac{\dot{m}_t^2 d_t}{\sigma\rho_m} \tag{4.37}$$

$$\rho_m = \left(\frac{x}{\rho_g}+\frac{1-x}{\rho_{sl}}\right)^{-1} \tag{4.38}$$

The correlation is applicable only if the ratio of slurry density to the gas density is less than 1000. If the model is applied, the coefficients must be changed according to the experimental observation in slurry reactor.

4.4.2.6   THEISSING MODEL

Theissing (1980) proposed a theory based on the interaction among the phases. He recommended an exclusive friction multiplier technique which can be represented as

$$\left(\frac{dp}{dz}\right)_f = \left[\left(\frac{dp}{dz}\right)_l^{1/(N\varphi)} (1-x)^{1/\varphi} + \left(\frac{dp}{dz}\right)_g^{1/(N\varphi)} x^{1/\varphi}\right]^{N\varphi} \tag{4.39}$$

where

$$\varphi = 3 - 2\left(\frac{2\sqrt{\rho_l/\rho_g}}{1+\rho_l/\rho_g}\right)^{0.7/n} \tag{4.40}$$

$$N = \frac{N_1 + N_2\left[\dfrac{(dp/dz)_g}{(dp/dz)_l}\right]^{0.1}}{1+\left[\dfrac{(dp/dz)_g}{(dp/dz)_l}\right]^{0.1}} \tag{4.41}$$

$$N_1 = \frac{\ln\left(\dfrac{(dp/dz)_l}{(dp/dz)_{l0}}\right)}{\ln(1-x)} \tag{4.42}$$

$$N_2 = \frac{\ln\left(\dfrac{(dp/dz)_g}{(dp/dz)_{g0}}\right)}{\ln(x)} \tag{4.43}$$

The models can be validated in the case of three-phase flow in the slurry bubble column reactor. It is noted that the coefficient depends on the experimental conditions and the physical properties of the system.

4.4.2.7   GHARAT AND JOSHI MODEL

Gharat and Joshi (1992) suggested a logical model to anticipate the pressure drop in homogeneous and heterogeneous flow patterns. As per their observation, the liquid-phase frictional pressure drop is directly related to the hold-up structure. They reported that the internal liquid circulation is an integral part of the flow pattern in the reactor. They have stated that an additional turbulence due to the bubble movement contributes to the frictional pressure drop. The overall pressure drop is the combined effect of the increased friction factor by turbulence and the velocity. The model can be represented in case of slurry–bubble column reactor by the following equation:

$$\Delta P_{f,tp} = \Delta P_{f,sp} + \Delta P_{f,AT} \tag{4.44}$$

$$\Delta P_{f,sp} = \frac{2 f_{f,sp} u_{sp}^2 \rho_{sl} \Delta z}{d_c} \tag{4.45}$$

$$\Delta P_{f,AT} = \frac{2 f_{f,AT} u_s^2 \rho_{sl} \Delta z}{d_c} \tag{4.46}$$

And then, the overall friction factor is expressed in terms of additional friction factor (f_{AT}) as

$$f_{tp} = \frac{f_{sp}}{\varepsilon_{sl}^2} + f_{AT} \tag{4.47}$$

The additional friction factor depends on the transverse component of turbulence intensity (u_y') of the flow and slip velocity of the bubble which is represented as

$$f_{AT} = 2 \left(\frac{u_y'}{u_s} \right)^2 \tag{4.48}$$

where

$$u_y' = \frac{1}{\sqrt{3}} \left[gl \left(u_{sg} - \frac{\varepsilon_g u_{sl}}{\varepsilon_{sl}} - \varepsilon_g u_s \right) \right]^{1/3} \tag{4.49}$$

According to Davies (1972), in the normal direction, the fluctuating velocity component practically equals the friction velocity. The parameter l is the degree of turbulence in single-phase pipe flow, which is given by

$$l = \chi d_c \tag{4.50}$$

where the parameter χ depends on the operating conditions which can be determined by Gharat and Joshi (1992) model as

$$\chi = 0.17 - 0.19 u_{sl} - 0.24\varepsilon_g + 0.015 u_g \qquad (4.51)$$

For the homogeneous flow,

$$u'_y = 1.5\varepsilon_g u_s \qquad (4.52)$$

where

$$u_s = \pm \frac{u_{sg}}{\varepsilon_g} \mp \frac{u_{sl}}{1-\varepsilon_g} \qquad (4.53)$$

where the positive sign is for the upflow and the negative for the downflow. The two-phase multiplier can then be expressed as

$$\phi_l^2 = \frac{\Delta P_{f,tp}}{\Delta P_{f,sp}} = \frac{\Delta P_{f,sp} + \Delta P_{f,AT}}{\Delta P_{f,sp}} \qquad (4.54)$$

Or

$$\phi_{sl}^2 = \frac{1}{\varepsilon_{sl}^2} + \frac{f_{AT}}{f_{sp}}\left(\frac{u_s}{u_{sl}}\right)^2 \qquad (4.55)$$

Or

$$\phi_{sl}^2 = \frac{1}{\varepsilon_{sl}^2} + \frac{2}{f_{sp}}\left(\frac{u'_y}{u_{sl}}\right)^2 \qquad (4.56)$$

The model can be applied to the gas-interacting downflow slurry reactor.

4.4.2.8 AWAD AND MUZYCHKA MODEL

Awad and Muzychka (2008) model can be expressed based on the ratio of three-phase frictional pressure gradient to the single-phase frictional pressure gradients of the slurry and gas flowing alone. According to the suggestion of Churchill (1977), the correlation can be used to calculate the fanning friction factor for the single-phase pressure drop. Their correlations for the slurry and gas multipliers can be expressed as

$$\phi_{sl,A-M} = \left[1 + \left(\frac{1}{X^2}\right)^q\right]^{1/q} \qquad (4.57)$$

and

$$\phi_{g,A\text{-}M} = \left[1+\left(X^2\right)^q\right]^{1/q}$$ (4.58)

The parameter X is to be evaluated by using Lockhart and Martinelli (1949) correlation. The parameter q is to be determined from experimental data and depends on the size of the reactor. The friction multiplier can be expressed as

$$\phi^2_{sl,A\text{-}M} = \left(\frac{f_{tp}}{f_{slo}}\right)\left(1+x\frac{\Delta\rho}{\rho_g}\right)$$ (4.59)

The frictional pressure drop can also be expressed in terms of interfacial pressure drop. Awad and Muzychka (2010) stated that the total frictional pressure drop gradient can be expressed as the sum of the frictional pressure drop of each single phase and the interfacial pressure drop gradient which is represented as

$$\left(\frac{dp}{dz}\right)_{f,tp} = \left(\frac{dp}{dz}\right)_{sl} + \left(\frac{dp}{dz}\right)_{i} + \left(\frac{dp}{dz}\right)_{g}.$$ (4.60)

The liquid multiplier in Lockhart and Martinelli (1949) correlation can then be obtained in terms of interfacial friction multiplier $\phi_{sl,i}$, which can be represented in case of the three-phase flow as

$$\phi^2_{sl,S\text{-}M} = 1+\phi^2_{sl,i} +\frac{1}{X^2}$$ (4.61)

$$\phi^2_{sl,i} = \frac{C_{A\text{-}M}}{X^m}.$$ (4.62)

The values of $C_{A\text{-}M}$ and m can be obtained from the experimental data, which are dependent on the flow behaviour.

4.4.2.9   SUN AND MISHIMA MODEL

Sun and Mishima (2009) represented a new correlation based on the Chisholm parameter (C). According to them, the parameter (C) can be expressed as a function of the flow quality and superficial Reynolds number of the slurry and the gas phases. According to Sun and Mishima (2009) correlation, the slurry multiplier can be expressed as

$$\phi_{l,S-M}^2 = 1 + \frac{C_{S-M}}{X^{1.19}} + \frac{1}{X^2} \tag{4.63}$$

where

$$C_{S-M} = \lambda \left(\frac{Re_g}{Re_{sl}} \right)^a \left(\frac{1-x}{x} \right)^b . \tag{4.64}$$

The variables λ, a, b can be obtained from the experimental data.

4.4.3 ANALYSIS OF FRICTIONAL PRESSURE DROP BY MECHANISTIC MODEL

The frictional pressure drop can be analysed based on the dynamic interactions of the phases (Majumder, 2016), which is described as follows.

4.4.3.1 ENERGY BALANCE

The energy balance of three-phase flow at isothermal condition in the vertical slurry reactor for the gas and slurry phase can be written as

$$-\Delta P_{tp} \varepsilon_g A_t u_{sg} + g\Delta Z A_t \varepsilon_g \rho_g u_{sg} + E_g = 0 \tag{4.65}$$

$$-\Delta P_{tp} (1-\varepsilon_g) A_t u_{sl} + g\Delta Z A_t (1-\varepsilon_g)\rho_{sl} u_{sl} + E_{sl} = 0 \tag{4.66}$$

By adding Equations 4.65 and 4.66, one gets

$$-\Delta P_{tp}[(1-\varepsilon_g)u_{sl} + \varepsilon_g u_{sg}] + [(1-\varepsilon_g)\rho_{sl} u_{sl} + \varepsilon_g \rho_g u_{sg}]g\Delta Z + (E_{sl} + E_g)/A_t = 0. \tag{4.67}$$

By summing, the expression for $(E_{sl} + E_g)$ can be obtained as

$$\Delta P_{fsl} A_t (1-\varepsilon_g)u_{sl} + \Delta P_{fg} A_t \varepsilon_g u_{sg} + E_b + E_{slip} + E_w = (E_{sl} + E_g) \tag{4.68}$$

where ΔP_{sfl} is the frictional pressure loss due to the slurry flow and ΔP_{fg} is the frictional pressure loss due to the gas flow. E_b is the rate of energy loss due to the bubble formation; E_{slip} is the rate of energy loss due to slip of gas–slurry interface and E_w is the rate energy loss due to wetting of thin liquid layer with the solid wall. Substituting Equation 4.68 for $(E_{sl} + E_g)$ in Equation 4.67 yields

$$\Delta P_{tp}[L_s / \rho_l + G / \rho_g]$$
$$= [L_s + G]g\Delta Z + \Delta P_{fg}G / \rho_g + \Delta P_{fsl}L_s / \rho_{sl} + [E_b / A_t + E_{slip} / A_t + E_w / A_t] \quad (4.69)$$

where L_s and G are the mass flux of slurry and gas, respectively, which are defined as

$$L_s = (1 - \varepsilon_g)u_{sl}\rho_{sl} \qquad (4.70)$$

$$G = \varepsilon_g u_{sg}\rho_g . \qquad (4.71)$$

4.4.3.2 ENERGY DISSIPATION DUE TO SKIN FRICTION

A model can be articulated to determine the frictional losses due to each gas and slurry phase in three-phase flow based on the following assumptions:

(i) The friction factor for each phase in the three-phase flow is a constant multiple, α' of that if only the flow of single phase takes place in the reactor.

(ii) The area of contact of each phase with wall in three-phase flow is α'' times than that of the only flow of single phase in the reactor. In case of gas phase, the values of α' and α'' depend on the number of bubbles coming in contact with the wall of the slurry reactor.

From these assumptions and from a simple overall momentum balance one can write:

For the slurry phase,

$$\frac{\Delta P_{fsl}(\text{Cross-sectional area})_{tp}}{\Delta P_{fsl0}(\text{Cross-sectional area})_{sl0}} = \frac{(\text{Wall shear stress})_{tp}}{(\text{Wall shear stress})_{sl0}} \frac{(\text{Area of contact with wall})_{tp}}{(\text{Area of contact with wall})_{sl0}} \quad (4.72)$$

$$\Rightarrow \frac{\Delta P_{fsl} A_t (1 - \varepsilon_g)}{\Delta P_{fsl0} A_t} = \frac{(0.5 f \rho_{sl} u_{sl}^2)_{tp}}{(0.5 f \rho_{sl} u_{sl}^2)_{sl0}} \times \frac{(\text{Area of contact with wall})_{tp}}{(\text{Area of contact with wall})_{sl0}}$$
$$= \frac{f_{slg}}{f_{sl0}} \cdot \frac{(u_{sl}^2)_{tp}}{(u_{sl}^2)_{sl0}} \times \alpha_{sl}'' \qquad (4.73)$$

$$\Rightarrow \frac{\Delta P_{fsl} A_t (1-\varepsilon_g)}{\Delta P_{fsl0} A_t} = \alpha'_{sl} \cdot \frac{1}{(1-\varepsilon_g)^2} \cdot \alpha''_{sl} = \frac{\alpha_{sl}}{(1-\varepsilon_g)^2} \tag{4.74}$$

$$\Rightarrow \frac{\Delta P_{sfl}}{\Delta P_{fsl0}} = \frac{\alpha_{sl}}{(1-\varepsilon_g)^3} \tag{4.75}$$

where the subscripts '$sl0$' denote to slurry phase only, 'tp' is for to slurry–gas three-phase, 'sl' denotes to slurry actual velocity, $(u_{sl})_{tp} = (u_{sl})_{l0}/(1-\varepsilon_g)$, $\alpha_{sl} = \alpha'_{sl} \alpha''_{sl}$ and cross-sectional area for the flow of slurry in three-phase is equal to $(1-\varepsilon_g)$ times the cross-sectional area of the reactor. The parameter α'_{sl} is defined as f_{slg}/f_{sl0} and α''_{sl} is defined as (Area of contact) $_{tp}/$ (Area of contact) $_{sl0}$. ΔP_{sfl0} is the frictional pressure drop due to liquid when only liquid phase flows through the slurry reactor. It is assumed that the area of contact of liquid with wall in three-phase flow is the same as the area of contact with wall in single-phase liquid flow. Therefore, α'_{sl} can be interpreted as α'_l, which is defined as f_{slg}/f_{sl0}. Similarly, for the gas phase,

$$\frac{\Delta P_{fg}}{\Delta P_{fg0}} = \frac{\alpha_g}{\varepsilon_g^3} \tag{4.76}$$

where ΔP_{fg0} is the single gas-phase frictional pressure drop

$$\alpha_g = \alpha'_g \alpha''_g . \tag{4.77}$$

4.4.3.3 ENERGY DISSIPATION BY PHASE INTERACTION

The bubbles moving through the fluid experience pressure forces that affect their motion. The pressure force due to the phase interaction can be expressed as

$$\Delta P_{FD} = \frac{1}{8} C_d \rho_{sl} u_s^2 \tag{4.78}$$

where C_d is the drag coefficient, u_s is the slip velocity. If there is no slip between liquid and solid particle, the pressure is affected. The interaction force depends on the size and shape of the bubble and the nature of the gas–slurry interface. The drag coefficient and slip velocity can be calculated as

$$u_s = \left| \frac{u_{sl}}{1-\varepsilon_g} - \frac{u_{sg}}{\varepsilon_g} \right| \tag{4.79}$$

The rate of energy dissipation based on the slip of the phases is then calculated by

$$E_{slip} = \Delta P_{FD} Q_m = \frac{1}{8} C_d \rho_{sl} u_s^2 A_t (G/\rho_g + L_s/\rho_{sl}) \tag{4.80}$$

4.4.3.4   ENERGY DISSIPATION FOR BUBBLE FORMATION

The rate of energy dissipation for the formation of bubbles depends on the bubble size, surface tension and the mass velocity of gas. It can be calculated as

$$E_b = \left(\frac{G}{\rho_g} A_t \right) \frac{6\sigma}{d_{be}} \tag{4.81}$$

where d_{be} is the equivalent bubble diameter. It is expressed by comparing the same volume of sphere if the bubbles are not in regular shape.

4.4.3.5   ENERGY DISSIPATION BECAUSE OF WETTING

Energy is also dissipated due to the resistance of wettability of the liquid. The upflow hydrophobic liquids have less resistance to flow compared to hydrophilic liquids. There is no energy loss due to wettability of gas with the slurry reactor wall because there will be no contact between the wall and the gas due to the existence of the thin layer of liquid between the bubble surface and the wall of the reactor. The rate of energy loss due to wettability of liquid with the wall can be represented as

$$E_w = \frac{\pi d_t u_{sl} \sigma_{sl}}{1-\varepsilon_g} \tag{4.82}$$

Wetting energy dissipation depends on the dynamic contact angle (θ) between liquid and solid wall. It is a function of capillary number (Ca) and related approximately as $\theta = Ca^{1/3}$, where the capillary number is defined as, $Ca = \mu_{sl} u_s / \sigma_l$. The solid is wet by the liquid if the contact angle is less than 90°, whereas if contact angle is greater than 90°, it is called non-wetting.; $\theta = 0$ indicates complete wetting and if $\theta = 180°$, it is complete non-wetting.

4.4.3.6 DETERMINATION OF MODEL PARAMETERS

The slurry- and gas-phase frictional pressure drop can be calculated from Equations 4.75 and 4.76, respectively. The parameter α_g in Equation 4.77 depends on the bubbles frequency coming in contact with the wall of the reactor. Since there is a thin layer of liquid through the wall of the reactor, no bubbles are assumed to touch the wall. Hence, it is considered that the area of contact of gas with wall is negligible compared to the area of contact of gas in single gas-phase flow and hence, the value of α_g can be taken as zero. The Fanning's equation can be used to calculate the single liquid-phase pressure drop as:

$$\Delta P_{fsl0} = 2 f_{sl0} \, \rho_{sl} u_{sl}^2 L / d \tag{4.83}$$

For the gas phase, it is

$$\Delta P_{fg0} = 2 f_{g0} \, \rho_g u_{sg}^2 L / d \tag{4.84}$$

The friction factor can be calculated as

$$f_{s0} = \frac{16}{Re_n} \quad \text{for laminar flow } (Re_n < 2100) \tag{4.85}$$

$$f_{s0} = \frac{0.079}{n^5 (Re_n)^{10.5^n}} \quad \text{for turbulent flow } (Re_n > 4000) \tag{4.86}$$

where Re_n is the Reynolds number based on the effective viscosity of non-Newtonian fluid flow since the slurry system behave as a non-Newtonian liquid. The Reynolds number is defined as

$$Re_n = \frac{d_c^n u_{sl}^{2-n} \rho}{8^{n-1} K} \left(\frac{4n}{3n+1} \right)^n \tag{4.87}$$

The flow within laminar and turbulent flow is called transition flow. In this transition flow, the friction factor f_{s0} can be calculated by the equation developed by Desouky and El-Emam (1989) as:

$$f_{s0} = 0.125\left[n^{\sqrt{n}}\left(0.0112 + \frac{1}{\mathrm{Re}_n^{0.3185}}\right)\right] \quad \text{for transition flow } (2100 < Re_n < 4000) \quad (4.88)$$

Substituting Equations 4.75 for ΔP_{sfl}, 4.76 for ΔP_{fg}, 4.80 for E_{slip}, 4.81 for E_b and 4.82 for E_w into Equation 4.69, one can get the final equation for δ_{tp} as

$$\delta_{tp} = (1-\psi_g)\left[\frac{1}{1-\psi_g} + \lambda\alpha_l + \phi + \kappa + \frac{\psi_g}{1-\psi_g}\Gamma\right] \quad (4.89)$$

In Equation 4.89, except α_l, all the parameters are known. α_l can be obtained from Equation 4.89 as

$$\alpha_l = \frac{1}{\lambda}\left[\frac{1}{1-\psi_g}(\delta_{tp} - \Gamma\psi_g - 1) - \phi - \kappa.\right] \quad (4.90)$$

where

$$\delta_{tp} = \frac{\Delta P_{tp}}{[\gamma_g\rho_g + (1-\gamma_g)\rho_{sl}]} \quad (4.91)$$

$$\gamma_g = \frac{G/\rho_g}{[G/\rho_g + L_s/\rho_{sl}]} \quad (4.92)$$

$$\lambda = \frac{2f_{sl0}u_{sl}^2 L_r}{g\Delta z(1-\varepsilon_g)^3 d_r} \quad (4.93)$$

$$\phi = \frac{(1/8)C_d u_s^2}{g\Delta z(1-\gamma_g)} \quad (4.94)$$

$$\Gamma = \frac{6\sigma}{\rho_g g\Delta z d_b} \quad (4.95)$$

$$\kappa = \frac{4\sigma/d_r}{\rho_{sl}g\Delta z(1-\varepsilon_g)^2} \quad (4.96)$$

$$\psi_g = \frac{\rho_g \gamma_g}{\rho_g \gamma_g + \rho_l (1 - \gamma_g)}$$ (4.97)

The α_l can be obtained from the experimental data and predicted by developing empirical correlations based on different operating variables and corresponding experimental data.

4.5 FRICTION FACTOR

4.5.1 GENERAL CORRELATION FOR FRICTION FACTOR

In the downflow gas-interacting slurry reactor, the friction factor depends on different operating variables (Sivaiah and Majumder, 2012). It varies with superficial slurry and gas velocities. The variation in friction factor with superficial slurry velocity is much higher compared to the superficial gas velocity. The friction factor decreases with an increase in slurry concentration. Also, as the slurry concentration increases, the particle mobility and velocity are reduced due to the decrease in the population of gas bubbles inside the column, which, in turn, decreases the friction in the column.

In the downflow gas-interacting slurry bubble column, it is advisable to interpret the complex phenomena by individual variable with certain physical law due to the complex system; however, it is difficult to predict the dependence variable based on the simultaneous action of different independent variables. Therefore, with the all independent variables, the formation of correlation provides the better prediction of multiphase phenomena. In this regard, in the downflow gas-interacting slurry reactor, the friction factors have been correlated by the dimensional analysis as the function of system variables, operating variables and geometric variables of the system as

$$f_{TPL} = f(u_{sl}, u_g, \mu_{sl}, \rho_{sl}, d_c, h_m, d_p, \sigma_l, g, d_j)$$ (4.103)

By using dimensional analysis, the functional relationship between the friction factor and various dimensionless groups can be represented as

$$f_{TPL} = K(\mathrm{Re}_{sl}, \mathrm{Re}_g, Mo, D_r)$$ (4.104)

The relationship of friction factor with dimensionless groups can be expressed as:

$$f_{TPL} = K \, \text{Re}_{sl}^a \, \text{Re}_g^b \, Mo^c \, D_r^d \qquad (4.105)$$

The functionality of Equation 4.105 can be estimated from the experimental data for the three regions of the flow (laminar, transition and turbulent) based on the Reynolds number of slurry by multiple regression analysis. Typical correlations developed by Sivaiah and Majumder (2013) in the DGISR within a wide range of operating conditions are as follows:

For $Re_{sl} < 2100$ (laminar region)

$$f_{TPL} = 3.378 \times 10^{-4} \, \text{Re}_{sl}^{-0.866} \, \text{Re}_g^{-0.069} \, Mo^{-0.495} \, D_r^{0.088} \qquad (4.106)$$

For $2100 < Re_{sl} < 4000$ (transition region)

$$f_{TPL} = 2.598 \times 10^{-4} \, \text{Re}_{sl}^{-0.851} \, \text{Re}_g^{-0.058} \, Mo^{-0.500} \, D_r^{0.090} \qquad (4.107)$$

For $4000 < Re_{sl}$ (turbulent region)

$$f_{TPL} = 4.144 \times 10^{-5} \, \text{Re}_{sl}^{-0.184} \, \text{Re}_g^{-0.058} \, Mo^{-0.333} \, D_r^{0.089} \qquad (4.108)$$

It is found that the developed correlations for three-phase friction factor (f_{tpl}) fits well with experimental values within the range of operating variables: $9.55 \times 10^2 < Re_{sl} < 49.03 \times 10^2$, $27.82 < Re_g < 83.47$, $1.04 \times 10^{-10} < Mo < 3.10 \times 10^{-10}$ and $0.52 \times 10^3 < Dr < 22.68 \times 10^3$.

4.5.2   ANALYSIS OF FRICTION FACTOR BASED ON VELOCITY DISTRIBUTION

The friction factor can also be estimated based on the experimental pressure drop or by the measured radial velocity distribution. The estimation of the friction factor by radial velocity profile is important in driving the heat and mass transfer correlation. The energy conservation equation using velocity distribution can be used to develop a correlation for friction factor as the functions of the Reynolds number. Taler (2016) developed a model based on the velocity profile to analyse the friction factor in tube which can be modified and used to interpret the friction factor in the downflow gas-interacting slurry bubble column. He reported that the Darcy–Weisbach friction factor f_D for the flow in tubes as

$$f_D = \frac{4\Delta \, Pr_w}{\rho L u_m^2} \qquad (4.109)$$

The velocity distribution and the friction factor can be obtained from the solution of the momentum conservation equation. The turbulent velocity profile is based on the time-averaged momentum conservation equation,

$$\frac{1}{r}\frac{d}{dr}\left[r\rho(v+v_t)\frac{d\bar{u}}{dr} \right] = \frac{dP}{dz} \qquad (4.110)$$

Equation 4.110 can be written in cylindrical coordinates with the following assumptions

$$\bar{u}_r = 0, \frac{d\bar{u}_r}{dr} = 0, \frac{d\bar{u}_r}{dz} = 0, \frac{\partial}{\partial z}\left(\overline{\rho u_z' u_r'} \right) = 0 \qquad (4.111)$$

where u_z' and u_r' are the longitudinal and radial fluctuating velocity components, respectively. The eddy diffusivity momentum is defined as

$$v_t = -\left(\overline{\rho u_z' u_r'} \right) / \left(\rho \frac{\partial \bar{u}_r}{\partial r} \right) \qquad (4.112)$$

Equation 4.112 can be evaluated based on the following boundary conditions

$$\bar{u}_z \big|_{r=r_w} = 0; \qquad \frac{d\bar{u}_z}{dr}\bigg|_{r=0} = 0 \qquad (4.113)$$

From the eddy viscosity and the velocity profile, the shear stress τ is defined as

$$\tau = -(\mu + \rho v_t)\frac{d\bar{u}_z}{dr} = -\rho(v+v_t)\frac{d\bar{u}_z}{dr} = -\mu(1+v_t/v)\frac{d\bar{u}_z}{dr} \qquad (4.114)$$

The momentum conservation equation for a control volume of a diameter $d_w = 2r_w$ and a length Δz as shown in Figure 4.4 yields

$$\tau_w = -\frac{r_w}{2}\frac{dP}{dz} \qquad (4.115)$$

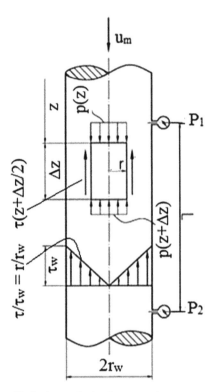

FIGURE 4.4   Flow profile in the vertical column reactor.

From Equations 4.114 and 4.115, Equation 4.112 can be rewritten as following:

$$\frac{1}{r}\frac{d}{dr}(r\tau) = \frac{2\tau_w}{r_w}$$

(4.116)

After the integration of Equation 4.116 with the boundary condition, $\tau|_{r=r_w} = \tau_w$, at $r = r_w$, it can be simplified as

$$\tau = \tau_w \frac{r}{r_w}$$

(4.117)

After the substitution of Equation 4.117, Equation 4.114 becomes

$$\frac{d\overline{u}_z}{dr} = -\frac{\tau_w r}{r_w \mu(1 + \nu_t / \nu)}$$

(4.118)

The integration of Equation 4.118 and solving Equation 4.118 with the boundary condition of Equation 4.113 yields the radial fluid velocity distribution. Based on the radial velocity profile $\bar{u}_z(r)$, the friction factor f_D can be determined. Defining the pressure gradient as

$$\frac{dP}{dz} = -\frac{f_D}{d_w}\frac{\rho u_m^2}{2} \tag{4.119}$$

and substituting Equation 4.119 into Equation 4.115 gives the expression for the shear stress at the wall as

$$\tau_w = \frac{1}{8}f_D\rho u_m^2 \tag{4.120}$$

where the mean velocity u_m is given by

$$u_m = \frac{2}{r_w^2}\int_0^{r_w}\bar{u}_x r\,dr \tag{4.121}$$

Announcing the so-called friction velocity u_* given by Lienhard and Lienhard (2011).

$$u^* = \sqrt{\tau_w/\rho} \tag{4.122}$$

The dimensionless variables can be formulated in the following way

$$y^* = \frac{(r_w - r)\sqrt{\tau_w/\rho}}{\nu} \tag{4.123}$$

$$r^* = \frac{ru^*}{\nu} = \frac{r\sqrt{\tau_w/\rho}}{\nu} \tag{4.124}$$

$$r_w^* = \frac{ru_*}{\nu} = \frac{\mathrm{Re}\sqrt{f_D/8}}{2} \tag{4.125}$$

$$R = r/r_w = 1 - y^*/r_w^* \tag{4.126}$$

Inserting Equation 4.122, for τ_w, into Equation 4.120 yields the friction factor for incompressible flow for fully developed flow (Lienhard and Lienhard, 2011).

$$f_D = \frac{8u_*^2}{u_m^2} = \frac{8}{(u_m^*)^2} \qquad (4.127)$$

$$u_m^* = u_m / u_* \qquad (4.128)$$

Equation 4.127 is used to calculate the friction factor for incompressible fully developed flow in a tube (Lienhard and Lienhard, 2011). The dimensionless velocity u^* can be determined experimentally. After that, the mean velocity u_m^* is calculated using Equations 4.121 and 4.128. For the details of the procedure to obtain the velocity distribution, the work published by Taler (2016) can be followed. The radial velocity profile can also be determined based on turbulent theory, which can be expressed as

$$u^* = \frac{1}{\kappa}\ln\left[(1+0.4y^*)\frac{1.5(1+R)}{1+2R^2}\right] + C\left[1-\exp\left(-\frac{y^*}{11}\right)-\frac{y^*}{11}\exp\left(-\frac{y^*}{3}\right)\right] \qquad (4.129)$$

where

$$C = 5.5 - (1/\kappa)\ln\kappa \qquad (4.130)$$

where κ is a constant of value, 0.4. Equation 4.129 is the velocity profile valid for the entire interval $0 \le r^* \le r_w^*$.

4.5.3 ANALYSIS OF FRICTIONAL DRAG COEFFICIENT BY ELECTROCHEMICAL METHOD

The frictional force can be evaluated by the velocity gradient of the liquid at the surface of solid-based ionic mass transfer by the electrochemical techniques. The velocity gradient influences the diffusional ionic mass transfer if any diffusion mass transfer is carried out in the system by electrochemical technique. The mass transfer can be carried out by using a microelectrode mounted flush with a tube wall. Reiss and Hanratty (1963) suggested a model to obtain the velocity gradient S_p in the vicinity of a wall of rectangular or cylindrical shape. This model was established in the cases where thickness of the concentration boundary layer is less than that of the hydrodynamic boundary layer as shown in Figure 4.5.

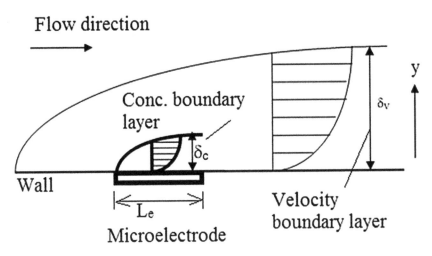

FIGURE 4.5 Hydrodynamic and concentration boundary layer.

The mass balance in the vicinity of the microelectrode can be written as follows:

$$\frac{\partial C}{\partial t} + S_p y \frac{\partial C}{\partial x} = D \frac{\partial^2 C}{\partial y^2}$$

(4.131)

The instantaneous rate of mass transfer is related to the concentration gradient as per Fick's law of diffusion at the surface over the electrode:

$$N = D \left\langle \frac{\partial C}{\partial y} \right\rangle$$

(4.132)

The mass transfer coefficient can be expressed as $k_L = N/(C - C_w)$, where C_w is the wall concentration. The analytical solution of Equation 4.131 yields the local mass transfer coefficient, which can be represented as

$$k_L = \alpha \left(\frac{D^2 S_p}{L_e} \right)^{1/3}$$

(4.133)

The method can be extended to spherical walls (sphere particle). A sphere is equipped with an inside channel, bent through 90°, in which a thread of 1 mm diameter is introduced, cut flush with the surface. A rigid tube in it serves as a support. The microelectrode can be directed relative to the average direction of the liquid by rotating the support as shown in Figure 4.6.

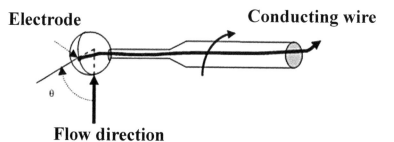

Electrode   **Conducting wire**

θ

Flow direction

FIGURE 4.6   Electrode to mount on the wall of reactor.

According to the principle of electrochemical technique, the apparent local mass transfer coefficient (and subsequently the velocity gradient at the wall) depends on the measurements of the diffusion—limiting current during the reduction of an electroactive species. The experimentation by the electrochemical process to estimate the friction force is given in detail in Chapter 7. The velocity gradient at the surface of the particle related to the apparent local mass transfer coefficient can be expressed as

$$S_p = (1.24 k_L)^3 \frac{d_e}{D_L^2}$$
(4.134)

where d_e is the electrode diameter and D_L is the diffusion coefficient of the ions. The values of diffusion coefficients can be evaluated by Equation 4.135 (Tonini et al., 1978) from the measured values of limiting current and superficial slurry velocity as

$$D_L = \left(\frac{i_L d_c L^{1/3}}{1.614 \, z \, F \, C_i \, A_e u_{sl}^{1/3} d_c^{2/3}} \right)^{3/2}$$
(4.135)

Then, the frictional drag force with Equation 4.134 can be estimated as

$$F_f = 2\pi R^2 \mu_{sl} S_p \int_0^{\pi/2} Sin^2 \theta \, d\theta$$
(4.136)

The frictional drag force on any spherical surface is characterized by the frictional drag coefficient (C_f) as (Lau et al., 2012)

$$C_f = \frac{F_f}{0.5\pi R^2 u_{sl}^2 \rho_{sl}}$$

(4.137)

where R is the radius of the sphere and F_f is the frictional drag force related to the velocity gradient at the surface of the particle (S_p). The frictional drag coefficient can be calculated from Equation 4.137 for different operating conditions by substituting the values of frictional drag force estimated by Equation 4.136. The coefficient increases with the fine size of particles. This is because the finer particles generate more friction as they try to adhere to the walls. A correlation for frictional drag coefficient as a function of slurry Reynolds number and slurry concentration developed by Sivaiah and Majumder (2013a) is given as

$$C_f = a_1 \, \text{Re}^2 + a_2 \, \text{Re} + a_3$$

(4.138)

where a_1, a_2 and a_3 vary with slurry concentration (w) as

$$a_1, a_2, a_3 = n_1 w^3 + n_2 w^2 + n_3 w + n_4$$

(4.136)

The values of the n_1, n_2, n_3, n_4 and regression coefficient (R^2) for different particle sizes are shown in Table 4.1.

TABLE 4.1 The Values of Parameters.

	$d_p = 2.21$ μm			$d_p = 19.39$ μm			$d_p = 96.0$ μm		
	a_1	a_2	a_3	a_1	a_2	a_3	$a_1 \times 10^3$	$a_2 \times 10^2$	$a_3 \times 10^2$
n_1	-7.27	1.55	-0.07	-0.004	0.000	-0.002	0.00	0.00	0.00
n_2	10.55	-2.24	0.10	0.008	-0.001	0.005	0.05	-0.03	0.05
n_3	-2.93	0.62	-0.03	-0.002	-0.001	-0.001	0.12	-0.14	0.37
n_4	1.16	-0.37	0.04	0.015	-0.042	0.037	0.37	-0.60	3.10
R^2	0.992	0.989	0.993	0.989	0.983	0.994	0.997	0.994	0.993

KEYWORDS

- **Awad and Muzychka model**
- **Bankoff model**
- **Baroczy model**
- **Chawla model**
- **energy**
- **energy balance**
- **frictional pressure drop**
- **Friedel model**
- **Gharat and Joshi model**
- **Lockhart–Martinelli correlation**
- **mass transfer**
- **Sun and Mishima model**
- **Theissing model**
- **Wallis model**

REFERENCES

Abdulrahman, M. W. Experimental Studies of Gas Holdup in a Slurry Bubble Column at High Gas Temperature of a Helium–Water–Alumina System. *Chem. Eng. Res. Des.* **2016,** *109,* (May) 486–494.

Abraham, J. P.; Tong, J. C. K.; Sparrow, E. M. Breakdown of Laminar Pipe Flow into Transitional Intermittency and Subsequent Attainment of Fully Developed Intermittent or Turbulent Flow. *Numer. Heat Transfer, Part A* **2008,** *54,* 103–115.

Awad, M. M.; Muzychka, Y. S. Effective Property Models for Homogeneous Two-Phase Flows. *Exp. Ther. Fluid Sci.* **2008,** *33,* 106–113.

Awad, M. M.; Muzychka, Y. S. Asymptotic Generalizations of the Lockhart–Martinelli Method for Two Phase Flows. *J. Fluids Eng.* **2010,** *132,* 1–12

Azbel, D.; Kemp-Pritchard, P. *Two-Phase Flows in Chemical Engineering*; Cambridge Univ. Press: New York, April 2009.

Bankoff, S. G. A Variable Density Single Fluid Model for Two Phase Flow with Particular Reference to Steam–Water Flow. *ASME J. Heat Trans. 82,* 265–272.

Barghi, S.; Prakash, A.; Margaritis, A. Flow Regime Identification in a Slurry Bubble Column from Gas Holdup and Pressure Fluctuations Analysis. *Can. J. Chem. Eng.* **2004,** *82,* 865–870.

Baroczy, C. J. A Systematic Correlation for Two-Phase Pressure Drop. *Chem. Eng. Prog.* 1966, *62*(64), 232–249.

Bukur, D. B.; Daly, J. G.; Patel, S. A. Application of γ-ray Attenuation for Measurement of Gas Holdups and Flow Regime Transitions in Bubble Columns. *Ind. Eng. Chem. Res.* **1996**, *35*, 70–80.

Chawla, J. M. Frictional Pressure Drop on Flow of Liquid/Gas Mixtures in Horizontal Pipes. *Chem. Eng. Tech.* **1972**, *44*, 58–63.

Chhabra, R. P.; Richardson, J. F. *Non-Newtonian Flow in the Process Industries. Fundamentals and Engineering Applications;* Butterworth-Heinemann: Oxford 1999; Vol. 6.

Chilekar, V. P.; Van der Schaaf, J.; Kuster, B. F. M.; Tinge, J. T.; Schouten, J. C. Influence of Elevated Pressure and Particle Lyophobicity on Hydrodynamics and Gas–Liquid Mass Transfer in Slurry Bubble Columns. *AIChE J.* **2010**, *56*, 584–596.

Chisholm, D. A. Theoretical Basis for the Lockhart-Martinelli Correlation for Three-phase Flow. *Int. J. Heat Mass Transfer* **1967**, *10*, 1767–1778.

Churchill, S. W. Friction Factor for Equations Spans all Fluid Flow Regimes. *Chem. Eng. J.* **1977**, *84*(24), 91–92.

Davies, J. T. *Turbulence Phenomena;* Academic Press: New York, 1972.

Davis, W. J. The Effect of Froude Number in Estimating Vertical Two Phase Gas-Slurry Friction Losses. *Br. Chem. Eng.* **1963**, *8*, 462–468.

Desouky, S. E. M.; El-Emam, N. A. Program Designs for Pseudoplastic Fluids. *Oil Gas J.* **1989**, *87*(51), 48–51.

Friedel, L. Pressure Drop During Gas/Vapour–Liquid Flow in Pipes. *Int. Chem. Eng.* **1980**, *20*, 352–367

Gharat, S. D.; Joshi, J. B. Transport Phenomena in Bubble Column Reactors II: Pressure Drop. *Chem. Eng. J.* **1992**, *48*(2), 153–166.

Gungor, A.; Eskin, N. Hydrodynamic Modeling of a Circulating Fluidized Bed. *Powder Technol.* **2006**, *172*, 1–13.

Han, L.; Al-Dahhan, M. H. Gas–Liquid Mass Transfer in a High Pressure Bubble Column Reactor with Different Sparger Designs. *Chem. Eng. Sci.* **2007**, *62*, 131–139.

Hatate, Y.; Nomura, H.; Fujita, T.; Tajiri, S.; Hidaka, N. A Ikari, Gas Holdup and Pressure Drop in Three Phase Vertical Flows of Gas-Liquid-Fine Solid Particles System. *J. Chem. Eng. Jpn.* **1986**, *19*, 56–61.

Heibel, A. K.; Vergeldt, F. J.; van As, H.; Kapteijn, F.; Moulijn, J. A.; Boger, T. Gas and Liquid Distribution in the Monolith Film Flow Reactor. *AIChE J.* **2003**, *49*, 3007–3017.

Holcombe, N. T.; Smith, D. N.; Knickle, H. N.; O'Dowd, W. Thermal Dispersion and Heat Transfer in Nonisothermal Bubble Columns. *Chem. Eng. Commun.* **1983**, *21*, 135–150.

Houzelot, J. L.; Thiebaut, M. F.; Charpentier, J. C.; Schiber, J. Contribution à l'étude hydrodynamique des colonnes à bulles. *Entropie (Paris)* **1983**, *19*, 121–126.

Iliuta, I.; Larachi, F.; Desvigne, D.; Anfray, J.; Dromard, N.; Schweich, D. Multicompartment Hydrodynamic Model for Slurry Bubble Columns. *Chem. Eng. Sci.* **2008**, *63*, 3379–3399.

Jordan, U.; Schumpe, A. The Gas Density Effect on Mass Transfer in Bubble Columns with Organic Liquids. *Chem. Eng. Sci.* **2001**, *56*, 6267–6272.

Joshi, J. B. Axial Mixing in Multiphase Contactors—A Unified Correlation. *Trans. Inst. Chem. Eng.* **1980**, *58*(1980), 155–165.

Joshi, J. B. Computational Flow Modeling and Design of Bubble Column Reactors. *Chem. Eng. Sci.* **2001**, *56*, 5893–5933.

Kiausner, J. F.; Chao, B. T.; Soo, S. L. An Improved Method for Simultaneous Determination of Frictional Pressure Drop and Vapor Volume Fraction in Vertical Flow Boiling Experimental Thermal and Fluid. *Science* **1990,** *3,* 404–415.

Krishna, R. A Scale-up Strategy for a Commercial Scale Bubble Column Slurry Reactor for Fischer–Tropsch synthesis. *Oil Gas Sci. Technol.—Rev. IFP* **2000,** *55,* 359–393.

Krishna, R.; Wilkinson, P. M.; Van Dierendonck, L. L. **1991.** A Model for Gas Holdup in Bubble Columns Incorporating the Influence of Gas Density on Flow Regime Transitions. *Chem Eng Sci.* 46(10), 2491–2496.

Kulkarni, A. A. Mass Transfer in Bubble Column Reactors: Effect of Bubble Size Distribution. *Ind. Eng. Chem. Res.* **2007,** *46,* 2205–2211.

Kunii, D.; Levenspiel, O.; Engineering, F. *Butterworth-Heinemann Series in Chemical Engineering,* 2nd ed.; Academic Press: New York, 1991.

Lau, R.; Lee, P. H. V.; Chen, T. Mass Transfer Studies in Shallow Bubble Column Reactors. *Chem. Eng. Proc.* **2012,** *62,* 18–25.

Leonard, C.; Ferrassea, J.-H.; Boutin, O.; Lefevre, S.; Viand, A. Bubble Column Reactors for High Pressures and High Temperatures Operation. *Chem. Eng. Res. Des.* **2015,** *100,* 391–421.

Li, W.; Zhong, W.; Jin, B.; Xiao, R.; He, T. Flow Regime Identification in a Three-Phase Bubble Column Based on Statistical, Hurst, Hilbert–Huang Transform and Shannon Entropy Analysis. *Chem. Eng. Sci.* **2013,** *102,* 474–485.

Lienhard, J. H., V.; Lienhard, J. H., IV. *Heat Transfer,* 4th ed.; Dover Publications: Mineola-New York, **2011.**

Lin, T. J.; Juang, R. C.; Chen, Y. C.; Chen, C. C. Predictions of Flow Transitions in a Bubble Column by Chaotic Time Series Analysis of Pressure Fluctuation Signals. *Chem. Eng. Sci.* **2001,** *56*(3), 1057–1065.

Lockhart, R. W.; Martinelli, R. C. Proposed Correlation of Data for Isothermal Two-Phase, Two Component Flow in Pipes. *Chem. Eng. Prog.* **1949,** *45,* 39–48.

Luo, X.; Lee, D. J.; Lau, R.; Yang, G. Q.; Fan, L. S. Maximum Stable Bubble Size and Gas Holdup in High Pressure Slurry Bubble Columns. *AIChE J.* **1999,** *45,* 665–680.

Maalej, S.; Benadda, B.; Otterbein, M. Interfacial Area and Volumetric Mass Transfer Coefficient in a Bubble Reactor at Elevated Pressures. *Chem. Eng. Sci.* **2003,** *58,* 2365–2376.

Majumder, S. K. *Hydrodynamics and Transport Processes of Inverse Bubbly Flow,* 1st ed.; Elsevier: Amsterdam, 2016, p 192.

Marchese, M. M.; Uribe-Salas, A.; Finch, J. A. Measurement of Gas Holdup in a Three-Phase Concurrent Downflow Column. *Chem. Eng. Sci.* **1992,** *47,* 3475–3482.

McCabe, W. L.; Smith, J. C. *Unit Operations of Chemical Engineering,* 3rd ed.; McGraw-Hill: New York, **1976.**

Nedeltchev, S.; Shaikh, A.; Al-Dahhan, M. Flow Regime Identification in a Bubble Column Based on both Statistical and Chaotic Parameters Applied to Computed Tomography Data. *Chem. Eng. Technol.* **2006,** *29*(9), 1054–1060.

Ong, Y. K.; Bhatia, S. Hydrodynamic Study of Zeolite-Based Composite Cracking Catalyst in a Transport Riser Reactor. *Chem. Eng. Res. Des.* **2009,** *87,* 771–779.

Pjontek, D.; Parisien, V.; Macchi, A. Bubble Characteristics Measured Using a Monofibre Optical Probe in a Bubble Column and Freeboard Region Under High Gas Holdup Conditions. *Chem. Eng. Sci.* **2014,** *111,* 153–169.

Reilly, I. G.; Scott, D. S.; De Bruijn, T. J. W.; MacIntyre, D. The Role of Gas Phase Momentum in Determining Gas Holdup and Hydrodynamic Flow Regimes in Bubble Column Operations. *Can. J. Chem. Eng.* **1994,** *72*(1), 3–12.

Reiss, L. P.; Hanratty, T. J. An Experimental Study of the Unsteady Nature of the Viscous Sublayer. *AIChE J.* **1963,** *9*(2), 154–160.

Roy, S.; Al-Dahhan, M. Flow Distribution Characteristics of a Gas–Liquid Monolith Reactor. *Catal. Today* **2005,** *105,* 396–400.

Ruthiya, K. C.; Chilekar, V. P.; Warnier, M. J. F.; Schaaf, J. V. D.; Kuster, B. F. M.; Schouten, J. C. Detecting Regime Transitions in Slurry Bubble Columns Using Pressure Time Series. *AlChE J.* **2005,** *51,* 1951–1965.

Sani, R. L. *Downflow Boiling and Non-boiling Heat Transfer in a Uniformly Heated Tube;* Report UCRL-*9032,* Univ. California, Lawrence Radiation Lab.: California, 1960.

Satterfield, C. N.; Ozel, F. Some Characteristics of Two-Phase Flow in Monolithic Catalyst Structures. *Ind. Eng. Chem. Fundam.* **1977,** *16,* 61–67.

Shaikh, A.; Al-Dahhan, M. Characterization of the Hydrodynamic Flow Regime in Bubble Columns via Computed Tomography. *Flow Meas. Instrum.* **2005,** *16*(2–3), 91–98.

Sivaiah, M.; Majumder, S. K. Gas Holdup and Frictional Pressure Drop of Gas-Slurry-Solid Flow in a Modified Slurry Bubble Column. *Int. J. Chem. React. Eng.* **2012,** *10,* 1–29.

Sivaiah, M.; Majumder, S. K. Mass Transfer and Mixing in an Ejector-Induced Downflow Slurry Bubble Column. *Ind. Eng. Chem. Res.* **2013,** *52,* 12661–12671.

Sun, L.; Mishima, K. Evaluation Analysis of Prediction Methods for Two-Phase Flow Pressure Drop in Mini-Channels. *Int. J. Multiphase Flow* **2009,** *35*(1), 47–54.

Taler, D. Determining Velocity and Friction Factor for Turbulent Flow in Smooth Tubes. *Int. J. Therm. Sci.* **2016,** *105,* 109–122.

Theissing, P. A Generally Valid Method for Calculating Frictional Pressure Drop in Multiphase Flow. *Chem. Eng. Technol.* **1980,** *52,* 344–355

Therning, P.; Rasmuson, A. Liquid Dispersion, Gas Holdup and Frictional Pressure Drop in a Packed Bubble Column at Elevated Pressures. *Chem. Eng. J.* **2001,** *81,* 331–335.

Tonini, R. D.; Remorino, M. R.; Brea, F. M. Determination of Diffusion Coefficients with a Pipe Wall Electrode. *Electrochim. Acta* **1978,** *23,* 699–704.

Wallis, G. B. *One Dimensional Two-Phase Flow;* McGraw-Hill Company: New York, **1969.**

Wilkinson, P. M.; Haringa, H.; Stokman, F. P. A.; Van Dierendonck, L. L. Liquid Mixing in a Bubble Column Under Pressure. *Chem. Eng. Sci.* **1993,** *48,* 1785–1791.

Wilkinson, P.; Haringa, H.; Van Dierendonck, L. L. Mass Transfer and Bubble Size in a Bubble Column Under Pressure. *Chem. Eng. Sci.* **1994,** *49*(9), 1417–1427.

Xu, G.; Nomura, K.; Nakagawa, N.; Kato, K. Hydrodynamic Dependence on Riser Diameter for Different Particles in Circulating Fluidized Beds. *Powder Technol.* **1999,** *113,* 80–87.

Xu, M.; Huang, H.; Zhan, X.; Liu, H.; Ji, S.; Li, C. Pressure Drop and Liquid Hold-up in Multiphase Monolithic Reactor with Different Distributors. *Catal. Today* **2009,** *147,* Supplement, 132–137

Yang, G. Q.; Fan, L.-S. Axial Liquid Mixing in High-Pressure Bubble Columns. *AIChE J.* **2003,** *49,* 1995–2008.

Yang, W.; Wang, J.; Jin, Y. Mass Transfer Characteristics of Syngas Components in Slurry System at Industrial Conditions. *Chem. Eng. Technol.* **2001,** *24,* 651–657.

CHAPTER 5

BUBBLE SIZE DISTRIBUTION

CONTENTS

5.1   INTRODUCTION

In gas-interacting slurry reactor, gas is distributed as a dispersed phase of bubbles of different sizes and shapes. Its size and shape generally ranges from 30 µm to 6 mm depending on the operating conditions and energy distribution. Gas-interacting slurry columns applied in chemical processes are characterized by extremely complex fluid dynamics interactions between the gas and slurry phase. Their correct design, operation and scale-up depend on the information of the fluid dynamics at different scales, mainly the 'bubble-scale' (i.e. bubble size distributions and shapes, single bubble dynamics, collective bubble dynamics) and the 'reactors-scale' (i.e. flow patterns, mean residence time of the disperse phase, dynamics of mesoscale clusters) (Besagni et al., 2016). The fluid flow behaviour at different scales can be estimated through the precise quantification of the local (i.e. the bubble size distributions and the bubble aspect ratio) and the global (i.e. the gas hold-up) fluid dynamic properties. The importance of bubble size and its distribution is summarized in the following sections:

Importance of bubble size and its distribution

- To design and optimize gas-interacting slurry reactor it is essential to recognize the prevailing flow regime and, subsequently, estimate the local and the global fluid dynamic properties. In this respect, a complete knowledge of the global fluid dynamics of the bubble column relies on the complete knowledge of the 'bubble-scale'. (Besagni et al., 2016)
- Bubble size distribution affects liquid-phase velocity field and governs the formation of liquid or slurry circulation cell in the reactor.
- The global and local fluid dynamic properties are strictly related to the prevailing flow regime. The distinction between homogeneous flow regimes from the heterogeneous flow regime in the reactor depends on the bubble size distribution.
- The transition flow regime is identified by the appearance of the 'coalescence-induced' bubbles and is characterized by large flow macrostructures with large eddies and a widened bubble size distribution due to the onset of bubble coalescence (Besagni et al., 2016)
- The size and shape of the bubble–liquid interface in the dispersed phase characterize the heat and mass exchange.

- Modelling for reactive absorption of gas in solutions that use an axial dispersion model for the liquid phase and a gas-phase model can be developed based on the conservation of the bubble size distribution function incorporating bubble break-up and coalescence (Lage, 1999; Fleischer et al., 1996).

- The degree of interaction between the dispersed bubbles and uninterrupted liquid or solid, and the variation of interfacial morphology of dispersed bubbles depend on the bubble size.

- Bubble occupies a specified volume leaving some free space in the reactor depending on the bubble population and its size. The reduction of free space may increase the coalescence rate and, hence governs the different size distribution and, therefore, the effect on flow behaviour in the reactor.

- In mineral beneficiation by bubbling process as per flotation technique, particles exhibiting hydrophobic surfaces attach to the rising air bubbles. These bubble particle aggregates then rise to the top of the separating column forming a froth phase rich in the selected minerals. This phenomenon depends on the bubble size and its interfacial area. The interaction between rising bubbles and particles is at the heart of the flotation process. One way to improve the flotation process performance would certainly adjust the bubble size distribution with respect to the particle size distribution (Riquelme et al., 2016). To control the entire bubble size distribution, thus, becomes essential to investigate the relationship between the flotation kinetics and hydrodynamic characteristics of the process.

- The applicability of bubble science and technology in industrial processes results mainly from the following properties: large gas–liquid surface area; greater gas hold-up and slower rise velocity. It determines the residence time and, in combination with the gas hold-up, the interfacial area for the rate of interfacial heat and mass transfer (Ahmed et al., 2013).

- Smaller size of the bubbles increases gas–liquid interfacial area, which increases the rate of diffusion and, hence the gas hold-up. Smaller size also decreases the bubble rise velocity. It is, therefore, necessary to understand how different parameters affect the size distribution of bubbles in order to design an efficient bubble generator (Gordiychuk et al., 2016)

- Bubble size distribution and bubble velocity distribution are the key parameters in evaluating the drag forces on bubbles and in using the bubble population balance in computational fluid dynamics (CFD).
- An understanding of the physical mechanisms determining bubble size is crucial to any detailed theory of the transfer of heat, mass and momentum among phases, and is also necessary for the framing of the design of reduced-scale laboratory models to simulate bubble and droplet flows in industrial plant.
- Break-up and coalescence of bubbles and mass transfer are governing the local bubble-size distribution. The rates of break-up and coalescence depend on the local flow field. On the other hand, the local bubble-size distribution determines the momentum transfer and gets affected by the local flow field. There is the need for a large interfacial surface area in order to achieve the required mass transfer rates demanded by various chemical processes. For the processes, bubble size distribution is fundamental to develop a full understanding of process efficiency.

Gas-interacting slurry columns are used for many mass transfer processes. They are used in a variety of industrial applications ranging from stripping and absorption columns to three-phase slurry beds. Often the gas–liquid mass transfer is empirically determined from the time profile of the concentration of dispersed phase in the system where information about bubble size and its distribution and interfacial area are deserted. However, a number of suggestions have emphasized the implication of bubble characteristics in controlling the mass transfer in gas–liquid–solid reactor (Akita and Yoshida, 1974; Hebrard et al., 2001; Couvert et al., 1999).

5.2   ESTIMATION OF BUBBLE SIZE

An enormous number of measurement techniques, including intrusive and nonintrusive methods, have been developed to investigate the bubble size and its flow behaviour in three-phase flow systems (Boyer et al., 2002; Yang et al., 2007). Among the methods, the most used methods are high-speed photography (Barigou and Greaves, 1991; Manasseh et al., 2001; Alves et al., 2002; Majumder et al., 2006), electrical impedance tomography (Kim et al., 2002; Nissinen et al., 2014); capillary suction probes

(Barigou and Greaves, 1991; Laakkonen et al., 2005; Hasanen et al., 2006), optical wave-guided sensors (Rząąsa and Pląskowski, 2003; de Oliveira et al., 2015); endoscopic optical probes (Chabot et al., 1991); capacitance probes (Leifer et al., 2000; Warsito and Fan, 2001) and passive acoustic detection (Manasseh et al., 2001; Chanson and Manasseh, 2003). Optical and conductivity microprobes have been most often used to examine bubble properties (Cartellier, 1998; Mudde and Saito, 2001).

In the downflow gas-interacting slurry bubble column, the same method of Majumder et al. (2006) can be followed. The images can be taken at different axial positions for different operating conditions to study the axial variation of the bubble size. A suitable high-speed digital camera can be used to capture the images. During the image capturing, the gas–liquid–solid flow can be illuminated by a compact fluorescent lamp. The images should be taken at homogeneous bubbly flow regime for different operating variables. The images are to be extracted and processed to adjust the resolution and brightness by any image processing software (Sivaiah and Majumder, 2013). It is better at least around 100–120 bubbles to analyse the bubble size in each images. The bubble size in slurry bubble columns can also be estimated using a spectral analysis method applied to measured pressure time series (Chilekar et al., 2005). The pressure time series is measured in a slurry bubble column based on the local and global pressure fluctuations. The average bubble size is measured from the standard deviation of these local pressure fluctuations.

The global pressure fluctuations are measured instantaneously throughout the reactor and form the coherent part of the pressure time series. The local pressure fluctuations are caused by liquid velocity fluctuations and gas hold-up fluctuations. The global pressure fluctuations are caused by bubble formation, coalescence, break-up, eruption, oscillations of the gas–liquid suspension, and mechanical vibrations (Lamb, 1945). Davidson and Harrison (1963) suggested a correlation for the pressure fluctuation generated by a single moving bubble in an infinite medium, caused by the gas flow field around it in a fluidized bed. This correlation is translated to gas-interacting slurry reactor by Chilekar et al. (2005) and converted from its radial coordinates into the time domain for a two-dimensional (2D) and a 3D bubble, which can be represented as

$$P_b(t) \propto \rho_s g \frac{d_b^2}{u_b t} \qquad \text{for 2D bubble} \qquad (5.1)$$

$$P_b(t) \propto \rho_s g \frac{d_b^3}{u_b^2 t^2} \qquad \text{for 3D bubble} \qquad (5.2)$$

where d_b is the diameter of the bubble and u_b is the rise velocity of the bubble. According to Davies (1950), this rise velocity of the bubble is proportional to the square root of the bubble diameter and so

$$P_b(t) \cong d_b^{1.5} \qquad \text{for 2D bubble} \qquad (5.3)$$

$$P_b(t) \cong d_b^2 \qquad \text{for 3D bubble} \qquad (5.4)$$

Chilekar et al. (2005) reported that the liquid circulations in the slurry bubble reactor also generate local pressure fluctuations as a result of changes in the liquid velocity field influenced by its interaction with the rising bubbles.

Other several noninvasive techniques (Boyer et al., 2002) such as laser sheeting (Larachi et al., 1997), radioactive particle tracking (Dudukovic, 2002; Chen et al., 1999) and tomography (Schmitz and Mewes, 2000) are being used in gas-interacting slurry reactor. The noninvasive techniques are being used to estimate the bubble size distribution, bubble velocity distribution by video imaging techniques.

5.3   EQUIVALENT BUBBLE DIAMETER

All the bubbles obtained by the camera are not uniform in size. The shape of the bubble may be classified as spherical, oblate ellipsoidal and spherical/ellipsoidal cap and so forth as shown in Figure 5.1. The actual bubble shape depends on the relative magnitudes of the surface tension and inertial forces (Bhaga and Weber, 1981). For bubble diameter less than 1 mm in water, the surface tension force predominates and the bubble is approximately spherical.

For bubbles of intermediate size, the effects of both surface tension and the inertia of the medium flowing around the bubble are important and exhibit very complex shapes and motion characteristics. Ellipsoidal bubbles often lack fore- and aft-symmetry, and in extreme circumstances they cannot be described by any simple regular geometry due to significant shape fluctuations (Bhaga and Weber, 1981; Tsuge and Hibino, 1978).

Large bubbles, whose volumes are larger than 3 cm³, (i.e. $d_e > 18$ mm) in general, are dominated by inertial or buoyancy forces, with negligible effects of surface tension and viscosity of the liquid media. The bubble is approximately a spherical cap with an included angle of about 100° (provided liquid viscosity $\mu_l < 50$ mPa·s) and a relatively flat or sometimes indented base (Ref.: crelonweb.eec.wustl.edu). The fluctuation in the overall shape becomes suppressed, that is, the main feature of the bubble's shape fluctuation is the oscillation of the bubble base, especially in the edge region (Bhaga and Weber, 1981).

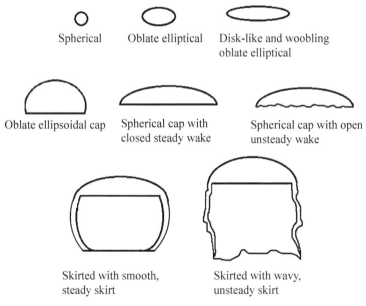

FIGURE 5.1 Different shapes of bubbles.

In multiphase flow systems, bubble size changes due to break-up and/or coalescence of bubbles caused by bubble–bubble interactions. The rates of these two processes are associated with the generated bubble size and the property of surrounding liquid and determination of interfacial area to govern the interfacial mass transfer and diffusion processes. Therefore, the equivalent mean bubble diameter is considered for further analysis especially for determination of interfacial area. A general form of mean diameter can be defined as

$$D_{pq} = \left[\frac{\int_0^\infty n(d_b)d_b^p \, dd_b}{\int_0^\infty n(d_b)d_b^q \, dd_b} \right]^{1/(p-q)} \tag{5.5}$$

Or in case of discrete bubbles

$$D_{pq} = \left[\frac{\sum_{i=1}^\infty n_i d_{bi}^p}{\sum_{i=1}^\infty n_i d_{bi}^q} \right]^{1/(p-q)} \tag{5.6}$$

where n_i is the number of bubbles of diameter d_{bi}. The mean diameter with the same ratio of volume to surface area is known as the Sauter mean diameter. The Sauter mean diameter is probably the most commonly used mean as it characterizes a number of important processes. In the above equations, it corresponds to values of $p=3$ and $q=2$. For most bubble size distributions, the Sauter mean diameter, D_{32}, is larger than the arithmetic, D_{10}, surface, D_{20}, and volume, D_{30}, mean diameters. After capturing the bubble image, if the size of the bubble is not spherical, whose maximum and minimum axes are known, an equivalent spherical bubble diameter can be calculated by the following equation (Couvert et al., 1999; Hebrard et al., 2001)

$$d_{be} = \sqrt[3]{l_{max}^2 l_{min}} \tag{5.7}$$

where l_{max} and l_{min} are the maximum and minimum axis length of a bubble. In fact, there may be different sizes of bubbles in the experimental test section. For that, it is necessary to represent the mean size of the bubble.

5.4   BUBBLE SIZE DISTRIBUTION AND ITS PREDICTION

The typical bubble size distribution with relative frequency which represents the relative fraction of the total counts for each interval (continuous random variable) or class (discrete random variable) at a constant slurry velocity (9.90×10^{-2} m/s) and slurry concentration (0.5%) is shown in Figure 5.2. The distributions of bubble size can be obtained by sorting the equivalent diameters of bubbles into different uniform classes. From the Figure 5.2, it is observed that the Pearson distribution (Eq. 5.8) fitted well compared to the other distributions (Log-Logistic, Lognormal, Beta and Weibull) (Sivaiah and Majumder, 2012).

$$f(x) = \frac{\beta^\alpha}{\Gamma(\alpha)(x-\min)^{\alpha+1}} \exp\left[-\frac{\beta}{(x-\min)}\right] \qquad (5.8)$$

where α = shape parameter > 0, β = scale parameter > 0 and min = minimum x. The values of α, β and *min* depend on the slurry velocity and the particle size.

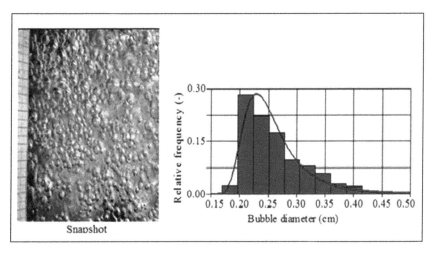

FIGURE 5.2 Typical bubble size distribution in gas-interacting downflow slurry bubble column.

Sivaiah and Majumder (2012) reported the Sauter mean bubble diameters for three particle diameters (2.21–96.0 μm) with different slurry concentrations in downflow gas-interacting slurry bubble column reactor. Tests were conducted at superficial solid–liquid slurry velocity of 7.10×10^{-2} to 14.2×10^{-2} m/s and superficial gas velocity of 0.85×10^{-2} to 2.55×10^{-2} m/s. The formation of bubbles in the homogeneous bubbly flow regime by the plunging liquid jet in the column is very complex phenomena, and there is no reliable model to predict the Sauter mean bubble diameter (D_{32}) in the gas–liquid–solid three-phase system. From their study, it was observed that the bubble size depends on different operating variables of the system (Sivaiah and Majumder, 2012).

Therefore, the Sauter mean bubble diameter (D_{32}) has been correlated by the dimensional analysis with the operating variables. According to the suggestion of Sivaiah and Majumder (2012), the Sauter mean bubble diameter (D_{32}) can be expressed as a function operating variables as

$$D_{32} = f(u_{sl}, u_g, \mu_{sl}, \rho_{sl}, d_p, \sigma_{sl}, g) \tag{5.9}$$

By applying the Buckingham's pi theorem of dimensional analysis leads to

$$D_{r,32} = f(U_r, \text{Re}_{slp}, Mo, Fr_p) \tag{5.10}$$

The correlation has been developed by multiple linear regression analysis which is expressed by the Equation 5.11.

$$D_{r,32} = 1.402 \times 10^{-15}\, U_r^{-0.092}\, \text{Re}_{slp}^{-0.738}\, Mo^{-1.748}\, Fr_p^{-0.026} \tag{5.11}$$

It was found that Equation 5.11 predicts the experimental values with correlation coefficient (R^2) of 0.992 and standard error of 0.085 to a good extinct within the range of operating variables: $39.3 < D_{r,32} < 715.8$, $2.78 < U_r < 16.67$, $0.086 < \text{Re}_{slp} < 9.41$, $1.04 \times 10^{-10} < Mo < 2.61 \times 10^{-10}$ and $5.31 < Fr_p < 925.63$.

The size distribution can be evaluated by other distribution models such as (i) exponential distribution: The distribution is a continuous bounded on the lower side. It can be expressed as (Majumder, 2016; Law and Kelton, 1991; Johnson et al., 1994).

$$f(x) = \frac{1}{\beta}\exp\left(-\frac{[x - \min]}{\beta}\right) \tag{5.12}$$

where, min=minimum x value, β =scale parameter=mean of x. (ii) Normal distribution: is an unbounded continuous distribution. It is defined as (Majumder, 2016; Johnson et al., 1994)

$$f(x) = \frac{1}{\sqrt{2\pi\sigma^2}}\exp\left(-\frac{[x - \mu]^2}{2\sigma^2}\right) \tag{5.13}$$

where μ=mean=$\sum_{i=1}^{N} x_i / N$, σ=standard deviation. The normal distribution is an unbounded continuous distribution. It is sometimes called a Gaussian distribution. (iii) Lognormal distribution: It is a continuous distribution bounded on the lower side. It is defined as

$$f(x) = \frac{1}{(x-\theta)\sqrt{2\pi\hat{\sigma}^2}} \exp\left(-\frac{[\ln(x-\theta)-\hat{\mu}]^2}{2\hat{\sigma}^2}\right) \tag{5.14}$$

$$\hat{\mu} = \sum_{i=1}^{n} \ln(x_i)/N \tag{5.15}$$

$$\hat{\sigma} = \sqrt{(\sum_{i=1}^{n}(\ln x_i - \hat{\mu}))/N} \tag{5.16}$$

where θ=minimum of x. $f(x)$ is always 0 at minimum x, rising to a peak that depends on both $\hat{\mu}$ and $\hat{\sigma}$, then decreasing monotonically for increasing x. By definition, the natural logarithm of a lognormal random variable is a normal random variable. The lognormal distribution can also be used to approximate the normal distribution, for σ, while maintaining its strictly positive values of x (Johnson et al., 1994). (iv) Gamma distribution: It is also a continuous distribution bounded on the lower side defined by (Majumder, 2016; Johnson et al., 1994)

$$F(x) = \frac{(x-\min)^{\alpha-1}}{\beta^\alpha \Gamma(\alpha)} \exp\left(-\frac{[x-\min]}{\beta}\right) \tag{5.17}$$

where min=minimum x, α=shape parameter>0, β=scale parameter>0. (v) Beta distribution: The beta distribution is a continuous distribution that has both upper and lower finite bounds. It can be used empirically to estimate the actual distribution before much data is available. It is expressed as (Majumder, 2016; Johnson et al., 1994)

$$f(x) = \frac{1}{B(p,q)} \frac{(x-\min)^{p-1}(\max-x)^{q-1}}{(\max-x)^{p+q-1}} \tag{5.18}$$

where $\min \leq x \leq \max$, min=minimum value of x, max=maximum value of x, p=lower shape parameter>0, q=upper shape parameter>0, B (p, q) is beta function. (vi) Weibull distribution: This is a continuous distribution bounded on the lower side (Weibull, 1951). It provides one of the limiting distributions for extreme values. The distribution is expressed as

$$f(x) = \frac{\alpha}{\beta}\left(\frac{x-\min}{\beta}\right)^{\alpha-1} \exp\left(-\left(\frac{[x-\min]}{\beta}\right)^{\alpha}\right) \tag{5.19}$$

where min = minimum x, α = shape parameter > 0, β = scale parameter > 0. Like the gamma distribution, it has three distinct regions. (vii) Pearson distribution: The Pearson distribution is a continuous distribution with a bound on the lower side. The distribution is expressed as (Majumder, 2016; Law and Kelton, 1991)

$$f(x) = \frac{\beta^{\alpha}}{\Gamma(\alpha)(x-\min)^{\alpha+1}} \exp\left(-\left(\frac{\beta}{[x-\min]}\right)\right) \tag{5.20}$$

where min = minimum x, α = shape parameter > 0, β = scale parameter > 0. (viii) Log-logistic distribution: The log-logistic distribution is a continuous distribution bounded on the lower side and is expressed by (Johnson et al., 1994) as

$$f(x) = \frac{p\left(\dfrac{x-\min}{\beta}\right)^{p-1}}{\beta\left[1+\left(\dfrac{x-\min}{\beta}\right)^{p}\right]^{2}} \tag{5.21}$$

where min = minimum x, p = shape parameter > 0, β = scale parameter > 0. The parameters for the included logistic distribution, L-α and L-β, are given in terms of the log-logistic parameters, LLp and $LL\beta$, by

$$L - \alpha = \ln(LL\beta) \tag{5.22}$$

$$L - \beta = 1/(LLp) \tag{5.23}$$

5.5 MAXIMUM STABLE BUBBLE DIAMETER

The size of bubbles produced within the mixing zone in the gas interacting downflow slurry bubble column as shown in Figure 1.1 in Chapter 1 can be determined by the forces acting on the bubble. In low viscous slurry, the bubbles are deformed by forces arising from velocity fluctuations acting over distances of the order of the bubble diameter, d_b. The reinstating force repelling the deformation of the bubble is due to surface tension acting at the gas–slurry interface. The ratio of these two forces is known as the Weber number which is expressed as

$$We = \rho_{sl}\overline{u_{sl}^2}d_b / \sigma \qquad (5.24)$$

where, $\overline{u_{sl}^2}$ is the average value of the squares of the velocity differences. The maximum stable bubble size can be calculated from Equation 5.25, which provides the slurry physical properties, critical Weber number and average of the squares of the velocity fluctuations acting over length scales of the bubble diameter.

$$d_{max} = \frac{We_c \sigma}{\rho_{sl}\overline{u_{sl}^2}} \qquad (5.25)$$

where, the average of the squares of the velocity fluctuations can be calculated as per Kolmogorov energy distribution law

$$\overline{u_{sl}^2} \propto \left(\frac{\epsilon\, d_b}{\rho_{sl}}\right)^{2/3} \qquad (5.26)$$

where the proportionality constant is ~2.0 (Batchelor, 1951), ϵ is the average energy dissipation rate per unit volume. The dissipation of energy depends on the turbulence of the bubble motion. Substituting Equation 5.26 into Equation 5.25 and rearranging, the maximum stable bubble diameter can be expressed as

$$d_{max} = 0.66 \times (We_c \sigma)^{0.6}\, \rho_{sl}^{-0.2}\, \epsilon^{-0.4} \qquad (5.27)$$

The high energy dissipation results in the generation of a number of very fine bubbles by break-up mechanism are carried downward by the bulk liquid motion. A critical Weber number can be chosen from a number of theoretical and numerical studies relating to the break-up of bubbles (Majumder, 2016). Ryskin and Leal (1984) reported that the critical Weber number was a function of the Reynolds number and the values lie within the range 0.95–2.76. In downflow bubble column with ejector system it is equal to 1.16 (Majumder et al., 2006).

5.6   ENERGY DISSIPATION BY BUBBLE MOTION

The energy depends on how the interfacial forces arise due to momentum transfer across the interface. In case of unsteady relative motion, it governs the drag force. The rising bubble in the bubble column experiences a lateral force when the flow pattern is non-uniform in the radial direction. Joshi (2001) described the details of the formulation of these forces. He reported that all the predicted flow patterns must satisfy the energy balance. The rate of energy supply from the gas phase to the liquid phase occurs by two different mechanisms (Dhotre and Joshi, 2004): (i) the rate of energy transfer from mean to turbulent flow, (ii) the bubble-generated turbulent energy, which takes part in momentum transfer. The rate of energy transfer from mean to turbulent flow can be expressed as

$$E_{mt} = \frac{\pi}{4} d_c^2 (\rho_{sl} - \rho_g) gh_m (1 - \varepsilon_g)(u_{sl} - u_b) \qquad (5.28)$$

The bubble-generated turbulent energy for momentum transfer is

$$E_{bt} = C_B \frac{\pi}{4} d_c^2 (\rho_{sl} - \rho_g) gh_m (1 - \varepsilon_g) u_b \qquad (5.29)$$

where C_B is the fraction of the bubble-generated turbulent energy. The range of C_B value is 0–1. The turbulence generated by gas bubbles does not take part in the momentum transfer of the liquid phase if $C_B=0$, whereas $C_B=1$ indicates the entire bubble generated turbulence completely contributes to the momentum transfer (Ekambara et al., 2005). The fraction of energy is dissipated to liquid motion. The other fraction $(1-C_B)$ is dissipated in the close vicinity of the gas–liquid interface and does not take part in the liquid phase transport phenomenon (Ekambara et al., 2005). The rate of the total energy transfer (E_T) is then represented by the addition of Equations 5.28 and 5.29 as

$$E_T = E_{mt} + E_{bt} = \frac{\pi}{4} d_c^2 (\rho_{sl} - \rho_g) gh_m (1 - \varepsilon_g)\{u_{sl} - u_b(1-C_B)\} \qquad (5.30)$$

In downflow gas-interacting slurry bubble column reactor the total energy supplied by the liquid jet is

$$E_T = \frac{1}{2} Q_{sl} \rho_{sl} u_{sl}^2 \qquad (5.31)$$

Equating Equations 5.30 and 5.31, one gets the fraction of gas-phase energy (C_B), which is transferred to the liquid phase turbulent kinetic energy. The fraction of gas phase energy depends on the slip velocity or bubble velocity relative to the liquid velocity (Sivaiah and Majumder, 2013).

As per the experimental results the following correlation has been developed to represent the variation of C_B with Froude number (Fr) within the range of variables: $0.43 < C_B < 0.86$ and $-0.33 < Fr < 2.13$

$$C_B = k\,\mathrm{Fr}^2 + m\,\mathrm{Fr} + n \qquad (5.32)$$

where k, m and n are the constants that vary with slurry concentration (w) which can be obtained from experimental value. The correlation for k, m and n are given as per experimental results as

$$k, m, n = a\,w^3 + b\,w^2 + cw + d \qquad (5.33)$$

$$
\begin{aligned}
k &= 0.18w - 0.404 \quad \text{for } d_p = 2.2\ \text{¼m} \\
&= 0.655w^3 - 2.146w^2 + 2.03w - 0.912 \quad \text{for } d_p = 19.39\ \mu m \qquad (5.33a) \\
&= 0.044w - 0.931 \quad \text{for } d_p = 96.0\,\mu m
\end{aligned}
$$

$$
\begin{aligned}
m &= -0.134w^2 + 0.062w + 0.608 \quad \text{for } d_p = 2.2\ \mu m \\
&= -0.189w^3 + 0.577w^2 - 0.530w + 0.797 \quad \text{for } d_p = 19.39\ \mu m \qquad (5.33b) \\
&= 0.037w + 0.613 \quad \text{for } d_p = 96.0\,\mu m
\end{aligned}
$$

$$
\begin{aligned}
n &= -0.041w + 0.658 \quad \text{for } d_p = 2.2\ \mu m \\
&= -0.045w^3 + 0.161w^2 - 0.185w + 0.672 \quad \text{for } d_p = 19.39\ \mu m \qquad (5.33c) \\
&= -0.038w + 0.668 \quad \text{for } d_p = 96.0\,\mu m
\end{aligned}
$$

At a certain axial position, bubble will balance its buoyancy force with the downward slurry momentum as Froude number tends to zero. At this condition, the fraction of the gas-phase energy is transferred to the liquid-phase, depending on the slurry concentration, particle diameter and physical properties of system. Typical range of C_B values at this stagnant condition is 0.658–0.617 for water—zinc oxide slurry of concentrations of 0.1–1.0%, respectively, whereas the range is 0.713–0.686 for glycerol—aluminium oxide slurry of the same range of slurry concentrations (Sivaiah and Majumder, 2013).

5.7   ANALYSIS OF AXIAL BUBBLE SIZE DISTRIBUTION

The size distribution model can be expressed based on a bubble number flux $\dot{N}(Z)$. The balance for the bubble number flux over a volume element can be written as (Atkinson et al., 2003)

$$\dot{N}(z_2) = \dot{N}(z_1) - \int_{z_1}^{z_2} P_c \dot{f}(z)dz \qquad (5.34)$$

where P_c is the probability that a collision between two bubbles will result in a coalescence event and $\dot{f}(z)$ is the rate of collision per unit volume. Statistical concerns analogous to the kinetic theory of gases in turbulent motions in the inertial subrange (Kamp et al., 2001; Majumder, 2016) can be used to obtain the collision frequency of bubbles. Using this approach an expression for the number of collisions of equal particles per unit time and volume can be derived (Kuboi et al., 1972). The number of collisions per unit time and volume of equal size bubbles is given by

$$\dot{f}(z) = \left(\frac{8\pi}{3}\right)^{0.5} n^2 d_b^2 u_t \qquad (5.35)$$

where n is the bubble number density, which for spherical bubbles is equal to

$$n = \frac{6\varepsilon_g}{\pi d_b^3} \qquad (5.36)$$

and u_t is the turbulent velocity which provides a measure of relative velocity of two bubbles in the liquid at a distance of bubble diameter, d_b (Hinze, 1975). According to Hinze (1975) the turbulent velocity is equal to

$$u_t = \left(\frac{\dot{E}_v d_b}{\rho_{sl}}\right)^{1/3} \qquad (5.37)$$

where $\dot{E}_v$ is the average energy dissipation rate per unit volume in the homogeneous bubbly zone. The rate of energy dissipation of turbulent kinetic energy per unit volume in the bubbly flow zone can be estimated by the following equation (Kamp et al., 2001)

$$\dot{E}_v = \frac{2\rho_l u_*^3}{\kappa d_c} \qquad (5.38)$$

The friction velocity (u_*) can be estimated from Blasius equation (Colin et al., 1991). κ is the von Karman constant, equal to 0.41. According to Colin et al. (1991), the friction velocity can be calculated for the bubbly pipe flow as follows

$$u_* = (u_{sl} + u_g)\sqrt{\frac{0.079\,Re_m^{-1/4}}{2}} \tag{5.39}$$

where Re_m is the mixture Reynolds number which is defined as

$$Re_m = \frac{\rho_l(u_{sl} + u_g)d_c}{\mu_{sl}} \tag{5.40}$$

Equations 5.34–5.40 can be used to interpret the axial bubble number flux in the ejector-induced inverse flow bubble column if the probability of coalescence, P_c is known. The coalescence occurs due to collision of two bubbles provided that the interaction time, t_i, between bubbles exceeds the drainage time t_d (Chesters, 1991). The drainage time is defined as the time required for drainage of the liquid film between bubbles to a critical rupture thickness. Kamp et al. (2001) reported the probability of coalescence in terms of the ratio of the drainage, t_d, and interaction times, t_i, for two bubbles. The ratio t_d/t_i can provide the first indication of whether coalescence will or will not occur under given condition (Kamp et al., 2001; Sivaiah and Majumder, 2013):

$$\begin{aligned} P_c(t_d/t_i) &= 0 \quad if \ t_d/t_i > 1 \\ &= 1 \quad if \ t_d/t_i < 1 \end{aligned} \tag{5.41}$$

where

$$\frac{t_d}{t_i} = k\frac{\rho_{sl}u_0 d^2/8\sigma}{\dfrac{\pi}{4}(\rho_{sl}C_{vm}d^3/3\sigma)^{1/2}} \tag{5.42}$$

In Equation 5.42, u_0 is the relative velocity of two bubbles at the onset of deformation, C_{vm} is the virtual mass coefficient normally taken to be a constant between 0.5 and 0.8 (Kamp et al., 2001). The Virtual mass coefficient C_{vm} describes the volume of displaced fluid that contributes to the effective mass of unit volume of the dispersed phase (Majumder, 2016). The virtual mass term in the momentum equations for dispersed flow represents the force essential to accelerate the mass of the surrounding continuous phase, in the close vicinity of a dispersed-phase fragment, such as a bubble or droplet, when the relative velocity of the phases changes. It is important only if the continuous-phase density is of the same order of magnitude as or much greater than that of the dispersed phase (Sivaiah and Majumder, 2013). The variation of C_{vm} depends on the ratio of the bubble diameter. For spherical equal-sized bubble diameter the value of C_{vm} is 0.785 (Kamp et al., 2001). The parameter, k in Equation 5.42 is a correction factor, which is a function of bubble size and interfacial properties, to take into account hydrodynamic effects such as bubbles bouncing of each other prior to coalescence occurring. Kamp et al. (2001) developed an explicit expression, involving an exponential function for the probability of coalescence as

$$P_c \propto \exp(-t_d / t_i) \qquad (5.43)$$

Sivaiah and Majumder (2013) reported that the main limitation with Equation 5.43 is that the mathematical definition for t_d does not include the effect of surface tension which can change the drainage time by orders of magnitude. To some degree, the constant K can be used to account for such effects. According to Atkinson et al. (2003), the drainage time can be replaced with a constant, K times the bubble persistence time t_p. The bubble persistence time is defined as the period that a bubble remains in the liquid before rupturing. It is a function of bubble size and liquid composition. Atkinson et al. (2003) developed a correlation for the persistence time (t_p) with air–water system and obtained as

$$t_p = 2.60 \times 10^7 d_b^{3.02} \qquad (5.44)$$

Therefore, Equation 5.43 can be written as

$$P_c = \lambda \exp(-K t_p / t_i) \qquad (5.45)$$

where, λ is the constant of proportionality. The probability of coalescence is determined by calculating the persistence and interaction times for the measured bubble size. λ and K are unknown parameters, to be adjusted to obtain a best fit for the bubble number flux ($\dot{N}(z)$) over a volume element inside the column calculated in accordance with Equation 5.34 (Sivaiah and Majumder, 2013). The profile of the bubble flux, $\dot{N}(z)$ in the bubbly low zone can be then be calculated as

$$\dot{N}(z) = \frac{24Q_g}{d_{32}^3 \times d_c^2}$$

(5.46)

The bubble size changes with increasing distance from the bottom of the column due to the coalescence of smaller bubbles. The change in bubble number over locations is the result of a change in bubble size due to coalescence. The probability of coalescence (P_c) depends on bubble–bubble interaction time (t_i) and bubble persistence time (t_p). The probability of coalescence increases with the increase in liquid flow rate. This is because of increasing collision frequency between bubbles with increase in liquid flow rate. Increase in kinetic energy due to increase in jet velocity results the increase in the turbulence intensity and bubble–bubble interaction and, hence the number bubble flux in the reactor.

5.8   BUBBLE EXCHANGE MODEL

Consider the slurry bubble column as shown in Figure 5.3 in which a slurry jet entrains gas in the dispersed gas of bubbles due to jet kinetic energy (Majumder, 2016). When bubbles are formed and dispersed by slurry jet, some bubbles will move inversely by momentum transfer of slurry jet and some bubbles will move upward due to their buoyancy effect. At certain cross section of a height some of the bubbles which are going upward and inverse direction may exchange between upward and inverse streams due to momentum variation. During the movement of dispersed phase the up and inverse flowing bubbles may interact with each other. As a result, bubbles may coalescence or break-up which creates different size of bubbles. Finer bubbles which are formed by break-up mechanism move inversely along with liquid and relatively higher size of bubbles which exhibit more buoyancy force than momentum of liquid move upward. At the same time, some bubbles may exchange the stream cross currently due to momentum change distribution. Let us consider that concentration (number density) of the bubbles flowing upward and inverse direction are n_u and n_d, respectively at a cross section of height z. Due to concentration difference among the two streams (upflowing and inverse flowing), there always exist some bubbles being exchanged between these two streams. The exchange is quantified by the exchange factor (E). Bubble dispersion has been taking place due to velocity distribution and because of turbulence which is accounted by the dispersion coefficient of bubble motion. A general number balance over a control volume will then be written as

$$(Out - In)_{bulk\ matter} + (Out - In)_{dispersed\ matter} + disappearance + accumulation = 0 \quad (5.47)$$

If the number balance is applied for upflowing bubble in a control volume which is at height z and a length of dz as shown in Figure 5.4, then

$$(Out - In)_{bulk\ matter} = (n_{u,z+dz} - n_{u,z})u_{sl}S_u \quad (5.48)$$

$$(Out - In)_{dispersed\ matter} = -S_u D_b \left(\left(\frac{\partial n_u}{\partial z} \right)_{z+dz} - \left(\frac{\partial n_u}{\partial z} \right)_z \right) \quad (5.49)$$

Disappearance due to exchange is $-EP(n_{u,z} - n_{d,z})dz$ where E is the parameter which signifies the exchange factor and accumulation is $S_u \left(\frac{\partial n_{u,z}}{\partial t} \right)dz$. Substituting different terms in Equation 5.47, one gets

$$S_u \left(\frac{\partial n_u}{\partial t} \right) = S_u D_b \left(\frac{\partial^2 n_u}{\partial z^2} \right) - S_u u_{sl} \left(\frac{\partial n_u}{\partial z} \right) + EP(n_{u,z} - n_{d,z}) \quad (5.50)$$

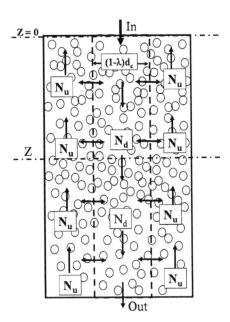

FIGURE 5.3   Schematic of bubble exchange.

Similarly, the number balance equation for inverse-flowing stream becomes

$$S_d\left(\frac{\partial n_d}{\partial t}\right) = S_d D_b\left(\frac{\partial^2 n_d}{\partial z^2}\right) + S_d u_{sl}\left(\frac{\partial n_d}{\partial z}\right) - EP\left(n_{u,z} - n_{d,z}\right) \qquad (5.51)$$

If there is a much interaction between the upward and inverse flowing dispersed phase, the exchange factor will be large; it can be assumed that the difference in dispersed phase concentration $(n_{u,z} - n_{d,z})$ will be relatively small. In other words, the radial concentration differences will then be small. Assume that the equivalent diameter of cross-sectional area through which bubbles move upward, d_u is equal to fraction λ of diameter of column (i.e. $d_u = \lambda d_c$). Then it can be expressed as

$$S_u = \lambda^2 S_c \qquad (5.52)$$

$$S_d = (1 - \lambda)^2 S_c \qquad (5.53)$$

Defining n as the volume average concentration at a cross section, it can be expressed as

$$N = \frac{n_u S_u \Delta z + n_d S_d \Delta z}{S_u \Delta z + S_d \Delta z} \qquad (5.54)$$

Or

$$N = \frac{n_u S_u + n_d S_d}{S_u + S_d}$$

By using Equations 5.52 and 5.53, Equation 5.54 can be simplified as

$$n = \frac{\lambda^2 n_u + (1-\lambda)^2 n_d}{1 - 2\lambda(1-\lambda)} \qquad (5.55)$$

This N signifies the total number of bubbles moving at a cross section. Defining concentration potential difference (M_d) as Equation 5.56 which signifies the difference in number of bubbles between the two passes (upflow (u) and inverse flow (d)) per unit volume

$$M_d = \lambda^2 n_u - (1-\lambda)^2 n_d \qquad (5.56)$$

Adding Equations 5.50 and 5.51 and further substitution of Equations 5.52 and 5.53, the resulting equation becomes

$$S_c\left(\frac{\partial n}{\partial t}\right) = S_c D_b\left(\frac{\partial^2 n}{\partial z^2}\right) - \frac{S_c u_{sl}}{1-\varepsilon_g}\left(\frac{\partial M_d}{\partial z}\right) \tag{5.57}$$

Subtracting Equation 5.51 from Equation 5.50 and using Equations 5.52 and 5.53, one gets

$$S_c\left(\frac{\partial M_d}{\partial t}\right) = S_c D_b\left(\frac{\partial^2 M_d}{\partial z^2}\right) - \frac{S_c u_{sl}}{1-\varepsilon_g}\left(\frac{\partial n}{\partial z}\right) - 2EP\left(n_u - n_d\right) \tag{5.58}$$

From Equations 5.55 and 5.56

$$n + M_d = 2\lambda^2 n_u \tag{5.59}$$

$$n - M_d = 2(1-\lambda^2)n_d \tag{5.60}$$

Equations 5.59 and 5.60 can be solved to get

$$n_u - n_d = \frac{(1-2\lambda^2)n + M_d}{2\lambda^2(1-\lambda^2)} \tag{5.61}$$

Substituting Equation 5.61 for $n_u - n_d$ in Equation 5.58 and further differentiating Equation 5.58 with respect to z, one can obtain

$$S_c\left(\frac{\partial^2 M_d}{\partial t \partial z}\right) = S_c D_b\left(\frac{\partial^2 M_d}{\partial z^2}\right) - \frac{S_c u_{sl}}{1-\varepsilon_g}\left(\frac{\partial^2 n}{\partial z^2}\right) - \frac{EP(1-2\lambda^2)}{\lambda^2(1-\lambda^2)}\left(\frac{\partial n}{\partial z}\right) - \frac{EP}{\lambda^2(1-\lambda^2)}\left(\frac{\partial M_d}{\partial z}\right) \tag{5.62}$$

By definition, M_d can be assumed to be a small value and, hence higher order derivatives of M_d can be neglected which modifies Equation 5.62 as

$$\frac{EP}{\lambda^2(1-\lambda^2)}\left(\frac{\partial M_d}{\partial z}\right) = -\frac{S_c u_{sl}}{1-\varepsilon_g}\left(\frac{\partial^2 n}{\partial z^2}\right) - \frac{EP(1-2\lambda^2)}{\lambda^2(1-\lambda^2)}\left(\frac{\partial n}{\partial z}\right) \tag{5.63}$$

Substituting $\partial M_d/\partial z$ from Equation 5.63 in Equation 5.57 and rearranging gives

$$\frac{\partial n}{\partial t} + \frac{u_{sl}(2\lambda^2-1)}{1-\varepsilon_g}\left(\frac{\partial n}{\partial z}\right) = \left[D_b + \left(\frac{u_{sl}}{1-\varepsilon_g}\right)^2\frac{\lambda^2(1-\lambda^2)S_c}{EP}\right]\frac{\partial^2 n}{\partial z^2} \tag{5.64}$$

Equation 5.64 forms the final model equation which represents a modified form of convection-dispersion model as

$$\frac{\partial n}{\partial t} + U\left(\frac{\partial n}{\partial z}\right) = \Gamma \frac{\partial^2 n}{\partial z^2} \qquad (5.65)$$

where

$$U = \frac{u_{sl}(2\lambda^2 - 1)}{1 - \varepsilon_g} \qquad (5.66)$$

$$\Gamma = D_b + \frac{\zeta}{E} \qquad (5.67)$$

where

$$\zeta = \left(\frac{\lambda^2(1 - \lambda^2)d_c}{4}\right)\left(\frac{u_{sl}}{1 - \varepsilon_g}\right)^2 \qquad (5.68)$$

Equation 5.64 represents modified dispersion model which expresses the overall dispersion coefficient of bubbles. At steady-state operation after integration of Equation 5.64 with following limiting conditions of Equations 5.69 and 5.70, the number density profile can be expressed by Equation 5.71 as follows:

$$\text{At} \quad z = z_f, \qquad n = n_0 \qquad (5.69)$$

$$\text{And} \quad z = L, \qquad n = \frac{\Gamma}{U}\left(\frac{dn}{dz}\right) \qquad (5.70)$$

$$n = n_0 \exp\left[\frac{U}{\Gamma}(z - z_f)\right] \qquad (5.71)$$

The profiles can be obtained by plotting n versus z. The parameter Γ can be calculated from the slope of the profile gradient. z_f is the distance from the top of the column where bubbles are started to form by gas entrainment due to plunging of liquid jet. The number density of bubbles which are formed at the formation zone at $z = z_f$ can be calculated as

$$n_{z=z_f} = n_0 = \frac{\text{Gas volume}}{\text{Bubble volume} \times \text{Bed volume}} = \frac{6\varepsilon_g}{\pi d_{b,e,0}^3} \qquad (5.72)$$

5.9   OTHER IMPORTANT POPULATION BALANCE MODELS

Gas–slurry flows are commonly encountered in many industrial flow systems. In many cases, the assessment of bubble size distribution is an important factor governing the momentum, heat and mass transfer between phases within the system. For the assessment of physical models and numerical schemes for slurry reactor, much effort is still required to validate the CFD tool due to the complex multiphase flow structures and its dynamical behaviour leading to transition of different flow regimes, an in-depth phenomenological understanding of bubble size or interfacial area and its dispersion behaviour is of paramount importance (Cheng et al., 2013). According to Cheng et al. (2013), bubbles within the bulk liquid flow undergo significant coalescence and break-up processes pose tremendous influence on the spectrum of bubble size. In the flow regime transition from bubbly flow, due to the rigorous bubble coalescence and break-up processes, the bubble size distribution gradually transforms from single-peaked to bi-modal or double-peaked profile. The bi-modal distribution signifies large amount of two types of bubbles (i.e. spherical and cap/distorted bubbles) that coexist within the system. Moreover, such change of bubble size or shape is also affected by lateral forces acting on the bubbles.

The evolution of bubble size distribution can be modelled by the population balance equation (PBE) which is expressed in integro-differential form with corresponding coalescence and break-up kernels. Various numerical models have been proposed to solve the PBE for practical systems (Ramakrishna, 2000). Among all existing methods, the 'class method' is widely adopted, in which the internal coordinate (e.g. particle length or volume) is discretized into a finite series of bins (Cheng et al., 2013). Multiple size group model (MUSIG) is encouraging for bubbly flow simulations. In the MUSIG model, the evolution of bubble size distribution is approximated by a number of discrete bubble classes each with individual transport equation and bubble coalescence and break-up kernels.

The homogeneous MUSIG model can be adapted to predict the population balance of gas–liquid–solid flows in various practical systems. The term homogeneous refers to the assumption that all bubbles travel in single velocity field. The model uses the computational resources for solving large number of transport equations which is excessive; especially for the flow condition in a large scale system where large bubbles could be prevalent. Cheung et al. (2013) proposed alternative simpler population balance model which is called the average bubble number density (ABND) model. They reported that encouraging results from the ABND model clearly demonstrated the capability of the model in capturing the required dynamical changes of bubble size due to coalescence and break-up.

Another inspiring model is also gaining considerable attention among researchers in the field to analyse the tracking of bubble population balance by the direct quadrature method of moments (DQMOM) that is developed by McGraw (1997). The DQMOM model directly resolves moments of the bubble size distribution with transport equations of abscissas and weighted abscissas which can be easily incorporated into generic CFD framework (Cheung et al., 2013). The main advantage of the DQMOM is its affability to endorse the change of bubble size by varying the abscissa of moments according to coalescence and break-up kernels. Instead of using a large number of bubble classes, DQMOM only requires tracking of a small number of moments.

The particle (bubble) size distribution according to the PBE can be expressed as (Fleischer et al., 1996):

$$\frac{\partial f(x,\xi,t)}{\partial t} + \nabla.(V(x,\xi,t)f(x,\xi,t)) = S(x,\xi,t) \tag{5.73}$$

where $f(x, \xi, t)$ is the particle (bubble) number density distribution per unit mixture and particle (bubble) volume, $V(x, \xi, t)$, is velocity vector in external space dependent on the external variables x for a given time t and the internal space ξ whose components is characteristic dimensions such as volume and mass. The term $S(x, \xi, t)$ contains the particle (bubble) source/sink rates per unit mixture volume due to the particle (bubble) interactions such as coalescence, break-up and phase change.

5.9.1   MULTIPLE SIZE GROUP (MUSIG) MODEL

The MUSIG model is the most commonly used technique for solving PBE (Lo, 1996). The MUSIG model considers that several bubble classes with different diameters d_j can be represented by an equivalent phase with the Sauter mean diameter (the volume–surface mean diameter). Generally, 10 bubble classes with the diameters ranging from minimum to maximum are considered based on the equal diameter discretization. The continuity equation for the dispersed phase, accounting for the changes in the particle size can be written as

$$\frac{\partial}{\partial t}\rho\alpha + \nabla.\rho u\alpha + \frac{\partial}{\partial r}\rho\dot{r}\alpha = 0 \qquad (5.74)$$

where t is time, ρ is density, α is volume fraction, u is velocity, r is particle radius and $\dot{r}$ is the rate of change in particle radius. The terms are signified as follows:

- First term: particle concentration variation with time,
- Second term: the convection in radial direction and
- Third term: the changes in the size of the group.

The continuity equation for size group-i can be written from the above equation as

$$\frac{\partial}{\partial t}\rho\alpha_i + \nabla.\rho u\alpha_i = S_i \qquad (5.75)$$

where S_i is the rate of mass transfer into or out of the size group due to break-up, coalescence, expansion, growth, and so forth. The sum of all the particle volume fractions equals the volume fraction of the dispersed phase (α)

$$\sum_i \alpha_i = \alpha \qquad (5.76)$$

The volume fraction of individual size-group is as:

$$f_i\alpha = \alpha_i \qquad (5.77)$$

Therefore, Equation 5.75 can be rewritten as

$$\frac{\partial}{\partial t}\rho\alpha f_i + \nabla.\rho u\alpha f_i = S_i \qquad (5.78)$$

where f_i is the fraction of the dispersed phase volume fraction in group-i. This equation has the form of the transport equation of a scalar variable, f_i, in the dispersed phase. Based on the transport equation PBE for bubble or drop as a dispersed phase can be written as

$$\frac{\partial(\rho_g\alpha_g f_i)}{\partial t} + \nabla(\rho_g\alpha_g f_i u_g) = S_i \qquad (5.79)$$

Or

$$\frac{\partial}{\partial t} n_i + \nabla . u n_i = B_B - D_B + B_C - D_C \tag{5.80}$$

where S_i is the source term because of bubble coalescence and break-up which is defined as:

$$S_i = \underbrace{B_{breakup} - D_{breakup}}_{\text{net source because of breakup to the ith group}} + \underbrace{B_{coalescence} - D_{coalescence}}_{\text{net source because of coalescence to the ith group}} \tag{5.81}$$

The net source because of break-up to the ith group is

$$B_i = \rho_g \alpha_g \left(\sum_{j>i} B_{ji} f_j - f_i \sum_{j<i} B_{ij} \right) \tag{5.82}$$

where the specific break-up rate of bubbles of ith group break into bubbles of jth group is denoted by B_{ij}, which is defined by (Jia et al., 2007)

$$B_{ij} = B'_{ij} \int_{f_{BV}} df_{BV} \tag{5.83}$$

where f_{BV} is the break-up fraction:

$$f_{BV} = \frac{m_j}{m_i} \tag{5.84}$$

where m_i is the mass of a particular group and can be related to the diameter of the group which is

$$m_i = \frac{\pi}{6} \rho_g d_i^3 \tag{5.85}$$

Based on the theory of isotropic turbulence and probability (Luo and Svendsen, 1996), the break-up rate of size i into bubbles of size j is

$$B'_{ij} = C_B (1 - \alpha_g) \left(\frac{\varepsilon}{d_i^2} \right)^{1/3} \int_{\varsigma_{min}}^{1} \frac{(1+\xi)^2}{\varsigma^{11/3}} \times \exp\left(-\frac{12(f_{BV}^{2/3} + (1 - f_{BV})^{2/3} - 1)\sigma}{\beta \rho 1 \varsigma^{2/3} d_i^5 \varsigma^{11/3}} \right) d\xi \tag{5.86}$$

where ε is the liquid phase eddy dissipation rate, and v is the kinematic viscosity. And the constants $C_B = 0.923$ and $\beta = 2$. The break-up rate depends on the surface tension σ and the dimensionless size of eddies in the inertial subrange of isotropic turbulence, ς. The minimum value of the dimensionless size of eddies in the inertial subrange of isotropic turbulence is

$$\xi_{min} = 11.4 \frac{1}{d_i}\left(\frac{1}{\varepsilon}v_1^3\right)^{1/4}$$
(5.87)

The net source to the ith group because of coalescence is (Jia et al., 2007)

$$C_i = (\rho_g \alpha_g)^2 \left(\frac{1}{2}\sum_{j \le i}\sum_{k \le i} C_{jk} f_j f_k \frac{m_j + m_k}{m_j m_k} X_{jki} - \sum_j C_{ij} f_i f_j \frac{1}{m_j} \right)$$
(5.88)

where C_{ij} is the specific coalescence rate between the ith and jth groups. X_{jki} is the fraction of mass because of coalescence between the jth and kth groups, which goes into group i (Jia et al., 2007).

$$X_{jki} = \begin{cases} \dfrac{m_j + m_k - m_{j-1}}{m_i - m_{i-1}} & m_{i-1} < m_j + m_k < m_i \\ \dfrac{m_{i+1} - (m_j + m_k)}{m_{i+1} - m_i} & if \ m_i < m_j + m_k < m_{i+1} \\ 0 & otherwise \end{cases}$$
(5.89)

The net source because of coalescence is zero when summed overall size groups which indicates

$$C_{ij} = C_{ji}$$
(5.90)

$$\sum_i X_{jki} = 1$$
(5.91)

Prince and Blanch (1990) described the coalescence process steps of two bubbles and summarized as follows:

Step 1: Collision of bubbles and trapping a small amount of liquid between them
Step 2: Draining of liquid film until the liquid film separating the bubbles reaches a critical thickness
Step 3: Rupturing the liquid film ruptures and joining the bubbles together

The coalescence process is, therefore, expressed by a collision rate of two bubbles. The collision efficiency is related to the times required for coalescence (t_{ij}) and the contact time (τ_{ij}). The collisions occur from three different contributions, turbulence, buoyancy and laminar shear (Prince and Blanch, 1990). The total coalescence rate is the summation of three contributions. Generally, buoyancy and laminar shear contribution are negligible compared to turbulence contribution. Based on the turbulent contribution, the expression for collision frequency is as follows (Jia et al., 2007):

$$C_{ij} = S_{ij} \ (u_{ti}^2 + u_{tj}^2)^{1/2} \eta_{ij} \qquad (5.92)$$

where the collision efficiency, η_{ij} is modelled by comparing the time required for coalescence t_{ij} with the actual time during the collision τ_{ij}.

$$\eta_{ij} = \exp(-t_{ij} / \tau_{ij}) \qquad (5.93)$$

$$t_{ij} = \left(\frac{\rho_1 r_{ij}^3}{16\sigma} \right)^{1/2} \ln \frac{h_o}{h_f} \qquad (5.94)$$

$$\tau_{ij} = \frac{r_{ij}^{2/3}}{\varepsilon^{1/3}} \qquad (5.95)$$

where h_o is the initial film thickness, h_f is the critical film thickness when rupture occurs, which are chosen as 1×10^{-4} and 1×10^{-8} m, respectively, and r_{ij} is the equivalent radius

$$r_{ij} = \left(\frac{1}{2} \left(\frac{1}{r_i} + \frac{1}{r_j} \right) \right)^{-1} \qquad (5.96)$$

The cross-sectional area of the colliding is defined by

$$S_{ij} = \frac{\pi}{4} (d_i + d_j)^2 \qquad (5.97)$$

The turbulent velocity is given by

$$u_{ti} = \sqrt{2} \ (\varepsilon d_i)^{1/3} \qquad (5.98)$$

The gas velocities by distributor are to be controlled as (Jia et al., 2007)

$$\frac{d_o^{1.5} u_g \ \rho_1 g^{0.5}}{\sigma} \leq 1 \qquad (5.99)$$

The size of bubbles produced at the distributor can be calculated by (Jia et al., 2007)

$$d_b = 2.9 \left(\frac{\sigma d_o}{g \rho_1} \right)^{1/3}$$

(5.100)

The boundary condition is to be considered as no-slip for the walls for the liquid phase and free-slip for the solid and the gas phases. A proper outlet condition is to be defined so that only gas phase can leave the domain through the inner area, and only liquid phase can leave the domain through the periphery, whereas solid phase is kept within the reactor (Jia et al., 2007).

5.9.2   AVERAGE BUBBLE NUMBER DENSITY (ABND) MODEL

According to Cheng et al., (2013), for isothermal bubbly flow, the population balance of dispersed bubbles can be expressed by an averaged quantity to quantify the overall changes of the bubble population. By integrating the PBE (Eq. 5.73), the ABND can be expressed as

$$\frac{\partial n}{\partial t} + \nabla.(u^g n) = R$$

(5.101)

where n is the ABND. The net rate of source and sink terms

$$R = (\varphi_n^{RC} + \varphi_n^{\Pi})$$

(5.102)

Equation 5.102 denotes the bubble number density variations due to random collision and turbulent-induced breakage. With the assumption of spherical bubbles in bubbly flow, the above transport equation is equivalent to the interfacial area transport equation (Hibiki and Ishii, 2002; Cheng et al., 2013). The phenomenological mechanism of coalescence and breakup source terms can be well understood given by Yao and Morel (2004). The rate of bubbles coalescence is

$$\varphi_n^{RC} = -C_{RC1} \frac{(\alpha^g)^2 (\varepsilon^l)^{1/3}}{D_s^{11/3}} \frac{\exp(-C_{RC2}\sqrt{We/We_{cr}})}{(\alpha_{max}^{1/3} - \alpha^g)/\alpha_{max}^{1/3} + C_{RC3}\alpha^g \sqrt{We/We_{cr}}}$$

(5.103)

The coefficients are $C_{RC1} = 2.86$, $C_{RC2} = 1.017$ and $C_{RC3} = 1.922$. Without coalescence caused by wake entrainment, the rate of bubble breakage is

$$\varphi_n^{\Pi} = C_{\Pi 1} \frac{\alpha^g (1-\alpha^g)(\varepsilon')^{1/3}}{D_s^{11/3}} \frac{\exp(-We_{cr}/We)}{1+C_{\Pi 2}(1-\alpha^g)\sqrt{We/We_{cr}}} \tag{5.104}$$

where the coefficients are $C_{\Pi 1} = 1.6$ and $C_{\Pi 2} = 0.42$. The equation is valid for maximum gas volume fraction (Hibiki and Ishii, 2002) of 50%.

5.9.3 DIRECT QUADRATURE METHOD OF MOMENTS (DQMOM)

The method of moments is derived based on the direct solution of the transport equations for weights and abscissas of the quadrature approximation (Fan et al., 2004; Cheng et al., 2013). It shares many similar features to QMOM (McGraw, 1997; Cheng et al., 2013). The model can be used to solve multidimensional problems. Marchisio and Fox (2005) developed QMOM by consideration of keeping track of the primitive variables appearing in the quadrature approximation instead of its moments where the abscissas and weights profiles are solved using matrix operations (Marchisio et al., 2006). The transport equations for calculating the weights and abscissas are written as

$$\frac{\partial n_i}{\partial t} + \nabla.(u^g n_i) = a_i \tag{5.105}$$

$$\frac{\partial \zeta_i}{\partial t} + \nabla.(u^g \zeta_i) = b_i \tag{5.106}$$

where $\zeta_i = n_i M_i$ is the weighted abscissas and the terms a_i and b_i are related to the birth and death rates which form 2N linear equations where the unknowns can be evaluated via Gaussian elimination (Cheng et al., 2013):

$$A\alpha = d \tag{5.107}$$

where the 2N vector of unknown α comprises essentially the terms a_i and b_i in Equations 5.105 and 5.106

$$\alpha = \left[a_1, \dots, a_N b_1, \dots, b_N \right]^T = \begin{bmatrix} a \\ b \end{bmatrix} \tag{5.108}$$

In Equation 5.107, the source or sink term on the right-hand side is defined by

$$d = [S_0, \ldots, S_{2N-1}]^T \qquad (5.109)$$

The moment transform of the coalescence and break-up of the term S_k for $k = 0, \ldots, 2N-1$ can be expressed as

$$S_k = (B_K^C - D_K^C + B_K^B - D_K^B) \qquad (5.110)$$

where the terms B and D represent the birth and death rates of the coalescence (B_K^C, D_K^C) and break-up (B_K^B, D_K^B) of bubbles, respectively. The weights and abscissas depend on the size fraction of the dispersed phase (f_k) and a variable defined as $\psi_k = f_k/M_k$. The size fraction of f_k can be expressed in terms of the weights and abscissas as

$$\rho^g \alpha^g f_k = n_i M_i = \zeta_k \qquad (5.111)$$

Based on Equation 5.111, the birth and death rates can be written as

$$B_K^C = \frac{1}{2} \sum_i \sum_j n_i n_j (M_i + M_j)^k a(M_i, M_j) \qquad (5.112)$$

$$D_K^C = \sum_i \sum_j M_i^k r(M_i, M_j) n_i n_j \qquad (5.113)$$

$$B_K^B = \sum_i \sum_j M_i^k b(M_j, M_i) n_j \qquad (5.114)$$

$$D_K^B = \sum_i \sum_j M_i^k r(M_i, M_j) n_i \qquad (5.115)$$

From above, the weights n_i and n_j can be resolved based on the definition given in Equation 5.111.

KEYWORDS

- **ABND model**
- **bubble exchange model**
- **bubble size**
- **bubble–bubble interaction**
- **bubble-scale**
- **DQMOM**

- **Gaussian distribution**
- **Lognormal distribution**
- **MUSIG model**
- **Weber number**

REFERENCES

Akita, K.; Yoshida F. Bubble Size, Interfacial Area and Liquid-Phase Mass Transfer Coefficient in Bubble Columns. *Ind. Eng. Process Des. Dev.* **1974,** *13*, 84–91.

Alves, S. S.; Maia, C. I.; Vasconcelos, J. M. T.; Serralheiro A. J. Bubble Size in Aerated Stirred Tanks. *Chem. Eng. Sci.* **2002,** *89*, 109–117.

Atkinson, B. W.; Gameson, G. J.; Anh, V. N.; Evans, G. M.; Machniewski P. M. Bubble Breakup and Coalescence in a Plunging Liquid Jet Bubble Column. *Can. J. Chem. Eng.* **2003,** *81*, 519–527.

Barigou, M.; Greaves M. A Capillary Suction Probe for Bubble Size Measurement. *Meas. Sci. Technol.* **1991,** *2*, 318–326.

Batchelor, G. K. *Proc. Cambridge Philos. Soc.* **1951,** *91*, 359.

Besagni, G.; Brazzale, P.; Fiocca, A.; Inzoli F. Estimation of Bubble Size Distributions and Shapes in Two-Phase Bubble Column Using Image Analysis and Optical Probes. *Flow Meas. Instrum.* **2016,** *52*, 190–207.

Bhaga, D.; Weber M. E. Bubbles in Viscous Liquids: Shapes, Wakes and Velocities. *J. Fluid. Mech.* **1981,** *105*, 61–85.

Boyer, C.; Duquenne, A. M.; Wild G. Measuring Techniques in Gas-Liquid and Gas-Liquid-Solid Reactors. *Chem. Eng. Sci.* **2002,** *57*, 3185–3215.

Cartellier, A.; Barraua, E. Monofiber Optical Probes for Gas Detection and Gas Velocity Measurements: Conical Probes. *Int. J. Multiphase Flow* **1998,** *24*, 1265–1294.

Chabot, J.; Lee, S. L. P.; Soria, A.; Lasa, H. I. Interaction Between Bubbles and Fiber Optic Probes in a Bubble Column. *Can. J. Chem. Eng.* **1991,** *70*, 61–68.

Chanson, H.; Manasseh, R. Air Entrainment Processes in a Circular Plunging Jet: Void-Fraction and Acoustic Measurements. *J. Fluids Eng. Trans. ASME* **2003,** *125*, 10–921.

Chen, J.; Kemoun, A.; Al-Dahhan, M. H.; Dudukovic, M. P.; Lee, D. J.; Fan, L. S. Comparative Hydrodynamics Study in Bubble Columns Using Computer automated Radioactive Particle Tracking (CARPT)/Computer Tomography (CT) and Particle Image Velocimetry. *Chem. Eng. Sci.* **1999,** *54*, 2199–2207.

Chesters, A. K. The Modelling of Coalescence Process in Fluid-Fluid Dispersions—A Review of Current Understanding. *Trans. I. Chem. E* **1991,** *69*(part A), 353–361.

Cheung, S. C. P.; Deju, L.; Yeoh, G. H.; Tu, J. Y. Modeling of Bubble Size Distribution in Isothermal Gas-Liquid Flows: Numerical Assessment of Population Balance Approaches. *Nucl. Eng. Des.* **2013,** *265*, 120–136.

Chilekar, V. P.; Warnier, M. J. F.; van der Schaaf, J.; Kuster, B. F. M.; Schouten, J. C.; van Ommen, J. R. Bubble Size Estimation in Slurry Bubble Columns from Pressure Fluctuations. *AIChE J.* **2005**, *51*(7), 1924–1937.

Colin, C.; Fabre, J.; Dukler, A. Gas Liquid Flow at Microgravity Conditions—I (Dispersed Bubble and Slug Flow). *Int. J. Multiphase Flow* **1991**, *17*, 533–544.

Couvert, A.; Roustan, M.; Chatellier, P. Two-Phase Hydrodynamic Study of a Rectangular Airlift Loop Reactor with an Internal Baffle. *Chem. Eng. Sci.* **1999**, *54*, 5245–5252.

Davidson, J. F.; Harrison, D. *The Exchange between the Bubble and Particulate Phases. Fluidised Particles;* Cambridge Univ. Press: Cambridge, UK, **1963**; p 67.

Davies, R. M.; Taylor, G.; Proc, R. *Soc. London Ser. A Phys. Sci.* **1950**, *200*, 375–390.

de Oliveira, W. R.; De Paula, I. B.; Martins, F. J. W. A.; Farias, P. S. C.; Azevedo, L. F. A. Bubble Characterization in Horizontal Air-Water Intermittent Flow. *Int. J. Multiphase Flow* **2015**, *69*, 18–30.

Dhotre, M. T.; Joshi, J. B. Two-Dimensional CFD Model for Pre-Diction of Pressure Drop and Heat Transfer Coefficient in Bubble Column Reactors. *Trans. IChemE Part A Chem. Eng. Res. Des.* **2004**, *78*, 689–707.

Dudukovic, M. P. Opaque Multiphase Flows: Experiments and Modeling. *Exp. Therm. Fluid Sci.* **2002**, *26*, 747–761.

Ekambara, K.; Dhotre, M. T.; Joshi, J. B. CFD Simulations of Bubble Column Reactors: 1D, 2D and 3D Approach. *Chem. Eng. Sci.* **2005**, *60*, 6733–6746.

Fan, R.; Marchisio, D. L.; Fox, R. O. Application of the Direct Quadrature Method of Moments to Polydisperse Gas-Solid Fluidized Beds, *Powder Technol.* **2004**, *139*(1), 7–20.

Fleischer, C.; Becker, S.; Eigenberger, G. Detailed Modeling of the Chemisorption of CO_2 into NaOH in a Bubble Column. *Chem. Eng. Sci.* **1996**, *51*, 1715–1724.

Gordiychuk, A.; Svanera, M.; Benini, S.; Poesio, P. Size Distribution and Sauter Mean Diameter of Micro Bubbles for a Venturi Type Bubble Generator. *Exp. Therm. Fluid Sci.* **2016**, *70*:51–60.

Hasanen, A.; Orivuori, P.; Aittamaa, J. Measurements of Local Bubble Size Distributions from Various Flexible Membrane Diffusers. *Chem. Eng. Process. Process Intensif.* **2006**, *45*, 291–302.

Hebrard, G.; Bouaifi, M.; Bastoul, D.; Roustan, M. A Comparative Study of Gas Hold-Up, Bubble Size, Interfacial Area and Mass Transfer Coefficients in Stirred Gas-Liquid Reactors and Bubble Columns. *Chem. Eng. Proc.* **2001**, *40*(2), 97–111.

Hibiki, T.; Ishii M. Development of One-Group Interfacial Area Transport Equation in Bubbly Flow Systems. *Int. J. Heat Mass Transfer* **2002**, *45*(11), 2351–2372.

Hinze, J. O. *Turbulence—An Introduction to its Mechanism and Theory;* McGraw-Hill: New York, NY, **1975**.

Jia, X.; Wen, J.; Zhou, H.; Feng, W.; Yuan, Q. Local Hydrodynamics Modeling of a Gas-Liquid-Solid Three-Phase Bubble Column. *AIChE J.* **2007**, *53*, 2221–2231.

Johnson, N. L.; Kotz, J. S.; Balakrishnan, N. *Continuous Univariate Distributions;* John Wiley and Sons: United States, 1994; Vol. 1, p 207.

Joshi, J. B. Computational Flow Modelling and Design of Bubble Column Reactors. *Chem. Eng. Sci.* **2001**, *56*, 5893–5933.

Kamp, A. M.; Chesters, A. K.; Colin, C.; Fabre, J. Bubble Coalescence in Turbulent Flows: A Mechanistic Model for Turbulence-Induced Coalescence Applied to Microgravity Bubbly Pipe Flow. *Int. J. Multiphase Flow* **2001**, *27*(8), 1363–1396.

Kim, M. C.; Kim, S.; Lee, H. J.; Lee, Y. J.; Kim, K. Y. An Experimental Study of Electrical Impedance Tomography for the Two-Phase Flow Visualization. *Int. Commun. Heat Mass Transf.* **2002**, *29*, 193–202.

Kuboi, R.; Komosawa, I.; Otake, T. Collision and Coalescence of Dispersed Drops in Turbulent Liquid Flow. *J. Chem. Eng. Japan* **1972**, *5*, 423–424.

Laakkonen, M.; Moilanen, P.; Miettinen, T.; Saari, K.; Honkanen, M.; Saarenrinne, P.; Aittamaa, J. Local Bubble Size Distributions in Agitated Vessel: Comparison of Three Experimental Techniques. *Chem. Eng. Res. Des.* **2005**, *83*, 50–58.

Lage, P. L. C. Conservation of Bubble Size Distribution During Gas Reactive Absorption in Bubble Column Reactors. *Braz. J. Chem. Eng.* **1999**, *16*(4), (São Paulo Dec.) http://dx.doi.org/10.1590/S0104-66321999000400002.

Lamb, H. *"Surface Waves." Hydrodynamics;* Cambridge Univ. Press/Macmillan: New York, NY, 1945; Ch. IX, p 364.

Larachi, F.; Chaouki, J.; Kennedy, G.; Dudukovic, M. P. Radioactive Particle Tracking in Multiphase Reactors: Principles and Applications. *Non-Invasive Monit. Multiphase Flows* **1997**, Chapter 11, 335–406.

Law, A. M.; Kelton, W. D. *Simulation Modeling and Analysis,* 5th ed.; McGraw-Hill: Michigan, 1991.

Leifer, I.; Leeuw, G., De; Cohen, L. H. Optical Measurement of Bubbles: System, Design and Application. *J. Atmos. Oceanic Technol.* **2000**, *17*, 392–1402.

Lo, S. M. Application of Population Balance to CFD Modelling of Bubbly Flow via the Musig Model. AEA Technology, AEAT-1096, 1996.

Luo, H.; Svendsen, H. F. Theoretical Model for Drop and Bubble Breakup in Turbulent Dispersions. *AIChE J.* **1996**, *42*(5), 1225–1233.

Majumder, S. K. *Hydrodynamics and Transport Processes of Inverse Bubbly Flow,* 1st ed.; Elsevier: Amsterdam, 2016.

Majumder, S. K.; Kundu, G.; Mukherjee, D. Bubble Size Distribution and Interfacial Phenomena in Ejector Induced Downflow Bubble Column. *Chem. Eng. Sci.* **2006**, *22*(1–2), 1–10.

Manasseh, R.; La Fontaine, R. F.; Davy, J.; Sheperd, I.; Zhu, Y. G. Passive Acoustic Bubble Sizing in Sparged Systems. *Exp. Fluids* **2001**, *30*(6), 672–682.

Marchisio, D. L.; Soos, M.; Sefcik, J.; Morbidelli M. Role of Turbulent Shear Rate Distribution in Aggregation and Breakage Processes. *AIChE J.* **2006**, *52*(1) (2006), 158–173.

McGraw R. Description of Aerosol Dynamics by the Quadrature Method of Moments. *Aerosol. Sci. Technol.* **1997**, *27*(2), 255–265.

Mudde, R. F.; Saito, T. Hydrodynamical Similarities Between Bubble Column and Bubbly Pipe Flow. *J. Fluid Mech.* **2001**, *437*, 203–228.

Nissinen, A.; Lehikoinen, A.; Mononen, M.; Lähteenmäki, S.; Vauhkonen, M. Estimation of the Bubble Size and Bubble Loading in a Flotation Froth Using Electrical Resistance Tomography. *Miner. Eng.* **2014**, *69*, 1–12.

Prince, M. J.; Blanch, H. W. Bubble Coalescence and Break-Up in Air-Sparged Bubblecolumns. *AICHE J.* **1990**, *36*(10), 1485–1499.

Ramakrishna, D. *Population Balances;* Academic Press: London, 2000.

Riquelme, A.; Desbiens, A.; del Villar, R.; Maldonado, M. Predictive Control of the Bubble Size Distribution in a Two-Phase Pilot Flotation Column. *Miner. Eng.* **2016**, *89*, 71–76.

Ryskin, G.; Leal, L. G. Numerical Solution of Free Boundary Problems in Fluid Mechanics Part 3. Bubble Deformation in an Axysymmetric Straining Flow. *J. Fluid Mech.* **1984,** *148,* 37–43.

Rzaąsa, M. R. Pląskowski, A.; Application of Optical Tomography for Measurement of Aeration Parameters in Large Water Tanks. *Meas. Sci. Technol.* **2003,** *14,* 199–204.

Schmitz, D.; Mewes, D. Tomographic Imaging of Transient Multiphase Flow in Bubble Columns. *Chem. Eng. J.* **2000,** *77,* 99–104.

Sivaiah, M.; Majumder, S. K. Gas Holdup and Frictional Pressure Drop of Gas-Liquid-Solid Flow in a Modified Slurry Bubble Column. *Int. J. Chem. React. Eng.* **2012,** *10*(1), 1542–6580 DOI: 10.1515/1542–6580.3035.

Sivaiah, M.; Majumder, S. K. Hydrodynamics and Mixing Characteristics in an Ejector-Induced Downflow Slurry Bubble Column [EIDSBC]. *Chem. Eng. J.* **2013,** *225,* 720–733.

Tsuge, H.; Hibino, S. Bubble Formation from a Submerged Single Orifice Accompanied by Pressure Fluctuations in a Gas Chamber. *J. Them. Eng. Jap.* **1978,** *11,* 173–178.

Warsito, W.; Fan, L. S. Measurement of Real-Time Flow Structures in Gas-Liquid and Gas-Liquid-Solid Flow Systems Using Electrical Capacitance, Tomography (ECT). *Chem. Eng. Sci.* **2001,** *56,* 6455–6462.

Weibull, W. A Statistical Distribution Function of Wide Applicability (PDF), *J. Appl. Mech.-Trans. ASME* **1951,** *18*(3), 293–297.

Yang, G. Q.; Du, B.; Fan, L. S. Bubble Formation and Dynamics in Gas-Liquid-Solid Fluidization-A Review. *Chem. Eng. Sci.* **2007,** *62,* 2–27.

Youssef, A. A.; Al-Dahhan, M. H.; Duduković, M. Bubble Columns with Internals: A Review. *Int. J. Chem. React. Eng.* **2013,** *11*(1), 169–223.

CHAPTER 6

DISPERSION PHENOMENA

CONTENTS

6.1   INTRODUCTION

Phase dispersion is one of the most important factors affecting the efficiency of the reactor to yield the efficient mass transfer operations. To determine the efficiency of the reactor, the information of the quality of dispersion is very important. For the scale-up and optimization, the dispersion mechanism is necessary. The understanding of phase dispersion is desired for the purpose of optimized operation and design of industrial-scale reactors. The investigation of dispersion based on local bubble motion would be beneficial for the optimization of operation, scale-up and design of the efficient slurry bubble column reactors and their applications. In many cases, several reactions occur simultaneously within a retention time of the dispersion process. The growth of the reactions, the product distribution and yield are governed by the dispersion of the components in the reactor. Axial dispersion coefficient is a design parameter that characterizes the intensity of dispersion in the reactor. As the dispersion is induced only by the gas aeration in the downflow gas-interacting reactor, a precise knowledge of the dispersion characteristics of the three phases in the gas-aided slurry reactor is required. The intensity of dispersion in a process unit can be characterized by using the residence time (t_m) of the phase. In order to improve the yield of various commercial chemical and bioprocesses, it is important to estimate the changes to the design and operation of the process unit that lead to a reduction in dispersion time. The factors that may affect the dispersion coefficient and dispersion time are column dimensions, gas distributor design, superficial gas velocity, the level of the fluid mixture inside the column and the physical properties of the fluid. This determines the overall rate of the reactions if the dispersion times are much larger than reaction times.

Various mathematical models have been suggested by several investigators to design and interpret the performance of the slurry bubble column reactors (Saxena et al., 1986; Turner and Mills, 1990). Most of the models are developed based on the studies of semi-batch operation in which the gas flows continuously through a batch liquid. Turner and Mills (1990) reported different types of developed dispersion models to interpret the performance of three-phase slurry bubble column reactors operated in semi-batch and continuous modes of operation. The semi-batch operation gives an intermediate level of performance to the cocurrent and countercurrent contactor; however, it may be preferred from a process flexibility perspective. They also reported that at low Peclet numbers, the approach to complete dispersion is so close that the efficiency of the reactor is indifferent to contacting pattern for liquid superficial velocities up to about 10% of the inlet gas superficial velocity.

Generally, the axial dispersion model (ADM) for the homogeneous flow patterns with interface mass transfer is suitable for design and modelling of the slurry reactors. The non-ideal models represent the simplest description of the flow pattern and do not account for the actual fluid dynamics of the system (Chen et al., 2006). The ADM is fitted with the experimental data obtained by residence time distribution (RTD) technique or conventionally called tracer technique. Chen et al. (2006) reported that the rapid scale-up and optimal commercialization of the reactor for the gas conversion processes operating under churn turbulent regime are needed, which is largely unknown and untested. It, therefore, necessitates an improved understanding and quantification of fluid dynamics and transport in the slurry reactor. The current representation of complex flow pattern in gas-interacting slurry reactor by the ADM stabs to swelling the explanation of too many physical phenomena into a single dispersion coefficient, which cannot be done in a precise manner (Degaleesan and Dudukovic, 1998). However, a verification of these models for reacting systems exposes them to the uncertainty of kinetics and additional proprietary constraints related to the revelation of kinetics (Chen et al., 2006). This may be one of the reasons why the performance analysis by various correlations differs usually. Second, the computational fluid dynamics codes may not be liable for the prediction of multiphase flow behaviour in the reactor until it is carefully validated with the experimental data, which may take long time. In the intervening period, it is necessary to develop engineering models for the depiction of dispersion and transport processes in the reactor. Dispersion characteristics, scale-up and design and various studies on the three-phase gas-interacting slurry reactors in details have been reported in the literature (Krishna and Sie, 2000; Degaleesan and Dudukovic, 1998; Gupta et al., 2001). Due to the industrial importance, majority of these studies reported on three-phase vertical upflow. The studies on dispersion mechanism of gas–liquid–solid three-phase cocurrent downflow slurry gas-aided slurry reactors are scanty.

6.2   ESTIMATION OF INTENSITY OF DISPERSION

Phase dispersion is the most important factor that affects the performance of the reactor to yield the efficient mass transfer operations (Majumder, 2008; Manish and Majumder, 2009). Therefore, to determine the efficiency of the gas-aided slurry reactors, the good knowledge of the quality of dispersion is very important. Axial dispersion coefficient is a design parameter that characterizes the intensity of dispersion. Axial dispersion coefficients in three-phase reactors are determined from the concentration profile of the tracer by using the steady-state tracer injection method (Tang and Fan, 1990). The time of the trace particle at a measuring point in the slurry gas-aided slurry reactor can be described by the ADM based on the assumptions reported by Majumder et al. (2005). The ADM can be expressed as

$$D_{sl} \frac{d^2 c_t}{dz^2} = \frac{dc_t}{dt} + u_{sl} \frac{dc_t}{dz} \tag{6.1}$$

where c_t is the tracer concentration measured at time t, u_{sl} is the actual slurry velocity, z is the axial distance between the point of tracer input and measuring point and D_{sl} is the axial dispersion coefficient of the slurry. The analytical solution for Equation 6.1 described by Levenspiel (1972) is

$$\frac{c_t}{c_o} = \frac{1}{2\sqrt{\pi \bar{t}/Pe}} e^{-\frac{(1-\bar{t})Pe}{4\bar{t}}} \tag{6.2}$$

where c_o is the input tracer concentration, Pe is the Peclet number defined as $u_{sl}z/D_{sl}$ and $\bar{t} = t_i/t_m$. In the axial dispersed plug flow, the residence time and the Peclet number can be determined by the moment method described by Levenspiel (1972). As per the moment method, the variance and mean of the concentration curve are related to the Peclet number and mean residence time, respectively, by

$$\sigma_\theta^2 = \frac{\sigma^2}{t_m^2} = \frac{8}{Pe^2} + \frac{2}{Pe} \tag{6.3}$$

where σ^2 is the variance and t_m is the mean residence time of the trace particle. The variance and the mean residence time can be calculated from the time-dependent experimental data of tracer concentration as follows:

$$\sigma^2 = \frac{\sum t_i^2 C_i \Delta t_i}{\sum C_i \Delta t_i} - t_m^2 \qquad (6.4)$$

$$t_m = \frac{\sum t_i C_i \Delta t_i}{\sum C_i \Delta t_i} \qquad (6.5)$$

6.3   RESIDENCE TIME DISTRIBUTION (RTD)

The intensity of the phase dispersion depends on different operating variables such as gas flow rate, liquid–solid slurry flow rate and slurry concentration (Sivaiah and Majumder, 2013). The effect of superficial gas velocity on the residence time of the tracer particle is shown in Figure 6.1. It is found that just after the injection of the tracer, up to a certain time the concentration of the tracer particles is zero in the sample collected. The reason is that when the tracer is injected, the particle of the tracer starts to disperse in the column and dispersion of the tracer particle takes places. As the tracer particle moves downward, more dispersion takes place, and in this process, it takes some time, so the samples collected below the column have almost zero concentration. The fraction of tracer particles' mean residence time (t_m) in the reactor is more (50.74 s) at lower gas velocities (0.85×10^{-2} m/s) compared to higher gas velocities (2.55×10^{-2} m/s) and the height of the peak of the curve is more at lower gas velocity. It indicates that the residence time decreases as the gas velocity is increased.

At higher gas velocity, the momentum energy in the gas–liquid–solid mixture increases and it enhances more dispersion, which increases the dispersion and reduces the residence time of the tracer particle. The effect of slurry velocity on the residence time of the tracer particle is shown in Figure 6.2. It is observed that the residence time of the tracer particle decreased with increasing superficial slurry velocity. At the higher slurry velocity, more energy dissipated for the gas–liquid–solid mixture and mixture gets more circulation, which enhances the dispersion of the liquid in the column. This may result in the variation of RTD of the tracer particle with superficial slurry velocity. At lower slurry flow rates, there will be a dispersion of liquid with a lower gas hold-up in the column. At this lower slurry velocity, the gas entrainment is relatively low and it does not produce internal circulation cell with gas–liquid transverse interactions. At higher slurry flow rates, there will be higher flow fluctuations, forming

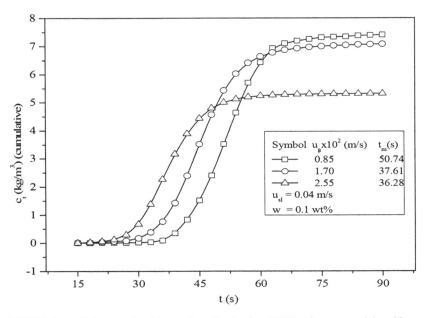

FIGURE 6.1   Variation of residence time distribution (RTD) of tracer particle with gas velocity for air–water–solids (Al_2O_3) system.

a longitudinal liquid slug and internal circulation of fluid in the column. This may vary the area under the curves in Figure 6.2. The curve peak reaches the maximum height and has higher mean residence time (36.40 s) at higher concentration of solids (1.0 wt.%) compared to the lower concentration of solids. This means that the residence time of the tracer particle is higher at higher slurry concentrations. As the slurry concentration increases, the downward movement of the liquid and the turbulence in the column decrease due to an increase in viscosity and the adherence of more particles to the bubble surface. The mean residence time of the tracer particle decreases with an increase in gas and slurry velocities. This is due to the increase in axial dispersion with the increase in gas velocity. When the superficial gas velocity in the downflow system is increased, the flowing gas bubbles tend to accelerate the liquid fluid element in the downward direction, so the velocity of the fluid element increases and hence the mean residence time of fluid element in the column decreases.

The mean residence time decreases with increasing superficial liquid–solid slurry velocity. This is due to the increase in turbulence in the column

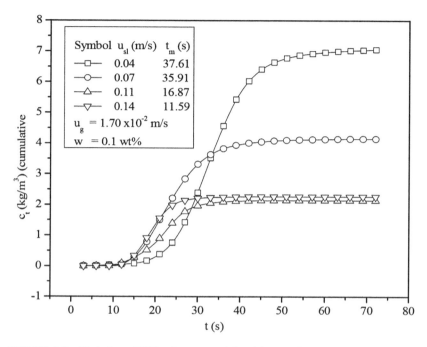

FIGURE 6.2 Variation of RTD of tracer particle with superficial slurry velocity for air–water–solids (Al_2O_3) system.

with increase in liquid flow rate. The turbulence in the column increases with an increase in the momentum exchange in the fluid element as the liquid–solid slurry flow rate increases. The high velocity fluid element is resulted and tries to disperse at a faster rate in the column, which leads to the reduction of the mean residence time. The variation in the mean residence time of the liquid is less at higher gas flow rates when compared to the lower gas flow rates. The reason is that at a higher gas flow rate, both the slurry and gas flow in the same direction with higher relative velocity, which in turn reduces the mean residence time of the tracer particle. The dispersion time is the mean residence time required to get a certain degree of dispersion. Rubio et al. (2004) expressed the dispersion time based on isotropic turbulence theory as

$$t_m = \beta \frac{z^2}{d_c^{4/3} \left(u_g g\right)^{1/3}}$$

(6.6)

where z is the dispersion height in the column, d_c is the column diameter and u_g is the actual gas velocity (u_g/ε_g). From the experimental data of gas-interacting downflow three-phase system, a relation is developed within the range of interstitial aeration velocity (u_g) from 0.02 to 0.85 m/s by the isotropic turbulence theory for the dispersion time, which can be represented as

$$t_m = \frac{(0.0602 + 0.0032w + 0.0114w^2)L^2}{d_c^{4/3}(u_g g)^{1/3}}\varepsilon_g - (0.7696 + 0.06w - 0.161w^2) \quad (6.7)$$

For the upflow slurry bubble column, the parameter β is found to be 0.44 (Rubio et al., 2004), whereas, in the downflow gas-interacting slurry column, the parameter is found to vary from 0.08 to 1.09 within the range of experimental conditions (Sivaiah and Majumder, 2013).

6.4   EFFECT OF DIFFERENT VARIABLES ON DISPERSION COEFFICIENT

From the literature, it is observed that the axial dispersion coefficient slightly decreases with increasing gas velocity in the bed of larger particles; however, in the bed of smaller particles, it increases with increasing gas velocity. The axial dispersion depends on the sizes of the columns. According to Kato and Nishiwaki (1972), the axial dispersion is directly proportional to the column diameter in the slurry bubble columns. They found that the dispersion coefficient depends on liquid velocity, gas velocity, particle size and the fluid properties. The axial and radial dispersion coefficients of the liquid phase in three-phase fluidized beds have been well represented by isotropic turbulence theory. Based on turbulence theory, the dispersion coefficient is increased with gas velocity and particle size; however, the dispersion coefficient exhibited a maximum value at a certain liquid velocity. The isotropic turbulence theory can be used to interpret the bubble-coalescing and bubble-disintegrating regimes. They observed that the axial dispersion coefficient increased with increasing liquid velocity, gas velocity, column size and particle size. They found that the effects of liquid surface tension and viscosity on dispersion coefficient are not significant. They also reported that the fluctuating frequency and the dispersion coefficient of particles increased with an increase in the gas velocity and particle size. They found that the fluctuating frequency and the dispersion coefficient decreased with an increase in liquid viscosity and increased with a decrease in surface tension of the liquid phase. Lu et al. (1994) found that the effects of solids loading on the intensity of dispersion in the riser was negligible, whereas, in the downcomer, it was increased initially and reached a maximum value and then decreased with increasing solids loading. They found that the axial dispersion coefficient increased with increasing aeration rate and static liquid height. The sparger designs also have a noticeable effect on gas hold-up and Peclet number at the lower gas velocity, that is, less than 0.025 m/s and at lower solid loading, that is, less than 2% (Cao et al., 2008). The turbulence in the downflow gas-interacting slurry column increases as the gas flow rate increases, which results in the deviation of the fluid more from plug flow and hence the dispersion coefficient increases. The effect of superficial gas velocity and slurry velocity on dispersion coefficient is shown in Figure 6.3. It is observed that on increasing the gas and slurry velocity, the dispersion coefficient increased. At the low value of superficial liquid–solid slurry velocity, the change in the value of dispersion coefficient with gas flow rate is small. An increase in the gas or slurry velocity in the column leads to an increase in the intensity of the turbulence, thereby increasing the dispersion of gas–liquid–solid mixture and enhancing the increase of the dispersion coefficient. Therefore, when the slurry flow rate is increased, the value of dispersion coefficient increases with gas flow rate and the variation in dispersion coefficient is also more at higher liquid flow rates. When the superficial gas velocity is increased in the downflow slurry bubble column, the flowing gas bubbles tend to accelerate the liquid fluid element in the downward direction; therefore, the velocity of the fluid element increases and leads to increase the dispersion of the liquid in the column.

The variation of slurry concentration is very less so the actual velocity of the liquid and liquid–solid is almost same in the column. The solid particle and the liquid both get the same intensity of dispersion in the column. As the gas hold-up increases, the flow area of the liquid in the column decreases; hence, the actual liquid velocity varies. This enhances the fluid circulation and formation of circulation cell in the column. This may increase the dispersion coefficient of the liquid in the column with slurry velocity. The dispersion coefficient decreases with an increase in the slurry concentration. An increase in the concentration of the solids, the

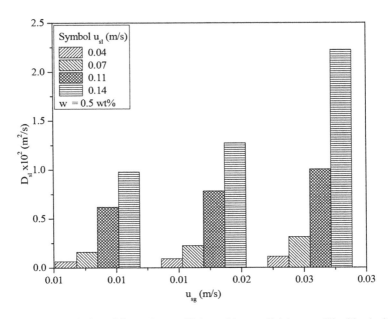

FIGURE 6.3   Variation of dispersion coefficient with superficial gas and liquid velocities.

viscosity of the slurry increases as shown in Figure 6.4. Consequently, the degree of dispersion of gas–liquid–solid mixture and the downward movement of the liquid and particles are hindered, which results in the increase of the residence time of the tracer particle and decrease the dispersion of liquid in the column. On increasing the slurry concentration, the particle mobility and velocity would be reduced due to the higher friction losses between the particle and the viscous liquid medium (Lim et al., 2011), which in turn decrease the dispersion coefficient. Also, as the slurry concentration increases, the breakup of the bubbles increases inside the column because of particle–particle interactions and it leads to the reduction of the gas hold-up and turbulence intensity in the column due to the formation of smaller bubbles in the column, thereby decreasing the degree of dispersion.

The axial liquid dispersion coefficient decreases with an increase in the separator pressure. As the separator pressure increases, the particles and the liquid face more resistance to moving downward due to their buoyancy effect and this causes more resistance to liquid dispersion. The separator pressure tends to decrease if the gas flow rate from the

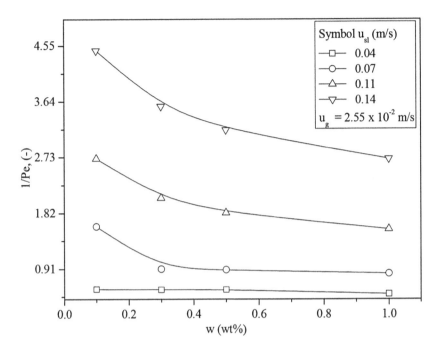

FIGURE 6.4 Variation of dispersion coefficient with slurry concentration for air–water–solids (Al_2O_3) system.

separator outlet increases and this results decrease of flow resistance inside the column and increases the axial dispersion coefficient. As the separator pressure increases, the pool resistance of the liquid–solid slurry inside the column increases, which results in a decrease in liquid dispersion inside the column. A comparison between the present experimental results of the axial liquid dispersion in ejector-induced gas–liquid–solid three-phase downflow gas-aided slurry reactor and the other published empirical correlations from the literature is shown in Table 6.1. From this comparison, it can be seen that for the same ranges of the operating parameters, the values of axial dispersion coefficient of downflow gas-interacting slurry reactor is higher than the other three-phase upflow studies. There is no pertinent evidence regarding the dispersion characteristics in the three-phase downflow gas-interacting slurry reactor. A comparison is shown in Table 6.1 with the dispersion intensity in three-phase upflow bubble column.

TABLE 6.1   Comparison of Axial Dispersion Coefficient with Different Other Published Correlations at u_g=0.025 m/s, Solid Concentration (w)=0.1 wt.% and u_{sl}=0.035–0.14 m/s at Normal Temperature And Pressure.

Reference	Range of parameters	System	Range of $E_z \times 10^2$ (m²/s)
Muroyama et al. (1978)	u_{sl}=0.006–0.168 m/s, u_g=0–0.3 m/s, d_c=0.06 and 0.1 m, d_p=0.215–6.9 (glass beads), 2.0 (alumina balls) and 5.2 mm (raschig rings)	Three-phase upflow	0.16–1.26
Kim and Kim (1983)	u_{sl}=0.02–0.13 m/s, u_g=0–0.12 m/s, d_c=0.145 m, d_p=1.7–6.0 mm, μ_{sl}=1–27 MPa·s, σ_{sl}=38.5–76 mN/m	Three-phase upflow	0.94–2.56
Kato et al. (1985)	u_g=0.02–0.16 m/s, d_c=0.012 m, d_p=0.52–5.2 mm, μ_{sl}=1.0–10.0 MPa·s	Three-phase upflow	0.30
Kang and Kim (1986)	u_{sl}=0.04–0.12 m/s, u_g=0–0.10 m/s, d_c=0.102 m, d_p=1.7–6.0 mm, μ_{sl}=1.0 MPa·s	Three-phase upflow	0.012–0.030
Kim et al. (1992)	u_{sl}=0.02–0.12 m/s, u_g=0.02–0.12 m/s, d_c=0.254–0.376 m, d_p=1.0–6.0 mm, μ_{sl}=1.0–72.5 MPa·s, σ_{sl}=32–72.8 mN/m	Three-phase upflow	0.92–2.49
Lee et al. (2007)	u_{sl}=0–0.03 m/s, u_g=0.001–0.008 m/s, d_c=0.152 m, d_p=4.0 mm, μ_{sl}=0.96–38 MPa·s, σ_{sl}=72.9–73.6 mN/m	Three-phase downflow	0.51–0.58
Lim et al. (2011)	u_{sl}=0.005–0.11 m/s, u_g=0.001–0.07 m/s, d_c=0.102 m, d_p=0.5–3.0 mm, μ_{sl}=1.0–38 mPa·s, σ_{sl}=52–72 mN/m	Three-phase upflow	0.046–0.051

TABLE 6.1   *(Continued)*

Reference	Range of parameters	System	Range of $E_z \times 10^2$ (m²/s)
Sivaiah and Majumder (2013)	u_{sl}=0.035–0.14 m/s, u_g=0.008–0.025 m/s, d_c=0.05 m, d_p=0.096 mm, μ_{sl}=1.4–1.6 MPa·s, σ_{sl}=74 mN/m	Three-phase downflow	0.15–4.03

The variation of axial dispersion coefficient with slurry velocity shows similar trend of increment as the other three-phase upflow studies, but the nature of the curve is different from the other works as a non-linear function of slurry velocity. In the downflow gas-interacting slurry bubble column, higher momentum exchange of plunging liquid jet enhance the fluid mixing. This may result in the dispersion coefficient of slurry phase (D_{sl}) in the downflow system as an order of magnitude is much higher than that in the upflow. The axial dispersion coefficient in the downflow gas-interacting slurry bubble column is a function of different operating variables such as gas velocity, slurry velocity, column diameter, particle diameter, density, viscosity and so on. Based on the experimental data, the axial dispersion coefficient has been well correlated in terms of the dimensionless groups of Reynolds number, Archimedes number and liquid-to-gas velocity ratios by the dimensional analysis, which can be expressed by Equation 6.8.

$$\frac{D_{sl}}{u_l z} = 6.64\times10^{-9} \ Re^{1.987} \ Ar^{0.594} V_r^{-0.604}$$

(6.8)

The correlation can predict the intensity of dispersion (dispersion number) within the range of operating variables: $11.22 \times 10^2 < Re < 96.11 \times 10^2$, $21.59 \times 10^{-2} < Ar < 27.37 \times 10^{-2}$ and $0.04 < V_r < 12.28$ to a good extinct.

6.5   ANALYSIS OF THE DISPERSION BY VELOCITY DISTRIBUTION MODEL

The dispersion fluid in the slurry reactor can be analysed by velocity distribution model based on fluid motion considering the gas interaction in the reactor. In the reactor, the liquid dispersion occurs under the action of the interaction of gas bubbles. At this condition, the dispersion coefficient of fluid can be represented by (Taylor, 1953; Aris, 1956)

$$D_{sl} = \frac{d_c^2 u_o^2}{kD_T} + D_T \qquad (6.9)$$

where u_o and D_T are the maximum velocity at the column axis and particle diffusion coefficient in either micro or macro level, respectively. The parameter K is the characteristic factor, which is a function of velocity distribution. The above model can be used to analyse the dispersion coefficient of bubble motion in downflow gas-interacting slurry bubble column (Sivaiah and Majumder, 2013) based on the assumptions: the flow is homogeneous in the column, uniform distribution of gas bubbles and the continuous dispersion of the solid and liquid throughout the column. Based on these assumptions, Equation 6.9 can be derived by replacing dispersion coefficient of bubble motion (D_b) in place of D_T as

$$D_{sl} = \frac{d_c^2 u_o^2}{KD_b} + D_b \qquad (6.10)$$

where D_b is the dispersion coefficient of the bubble motion and D_{sl} is the axial dispersion coefficient of the slurry, which is estimated by RTD method. The details of dispersion characteristics have been reported in the previous article by Sivaiah and Majumder (2013). In the downflow system, at steady state, the velocity of liquid at the column centre is twice the superficial slurry velocity (Majumder et al., 2005). Then, Equation 6.10 can be represented as

$$D_{sl} = \frac{4d_c^2 u_{sl}^2}{KD_b} + D_b \qquad (6.11)$$

The dispersion coefficient of bubble motion (D_b) and the characteristic factor (k) for different experimental conditions can be calculated from Equation 6.11 by plotting experimental data of axial dispersion coefficient of slurry (D_{sl}) versus the term $(4d_c^2 u_{sl}^2)$. An intercept of Equation 6.11 is equal to dispersion coefficient of the bubble motion (D_b) and the slope is equal to $1/(KD_b)$. It is observed that with an increase in the slurry concentration, the dispersion coefficient of the bubble motion (D_b) decreases and the characteristic factor (K) of velocity distribution increases. The interstitial slurry velocity inside the column varies with the overall gas hold-up.

This may affect the variation of D_b and K with slurry concentration and with superficial gas velocity. The correlations for dispersion coefficient of bubble motion (D_b) and velocity characteristic factor (K) as a function of superficial gas velocity (u_g) and slurry concentration (w) are developed within the range of superficial gas velocity from 0.85×10^{-2} to 2.5×10^{-2} m/s and slurry concentration of 0.1–1.0 wt.% and can be represented as

$$D_b = \alpha u_g^2 + \beta u_g + \gamma \qquad (6.12)$$

where

$$\alpha = 327.83w^3 - 553.41w^2 + 258.84w - 37.35 \qquad (6.13)$$

$$\beta = -12.29w^3 + 20.87w^2 - 9.85w + 1.43 \qquad (6.14)$$

$$\gamma = 0.087w^3 - 0.15w^2 + 0.069w - 0.0087 \qquad (6.15)$$

$$k = \left(-10.09w^2 + 10.81w - 0.68\right) u_g^{\left(-3.78w^2 + 3.61w - 1.53\right)} \qquad (6.16)$$

Majumder (2008) reported some characteristics of the parameters. The dispersion coefficient of bubble motion and characteristic factor of velocity distribution in up- and downward cocurrent flow system which can be summarized as follows:

- The characteristic factor of velocity distribution and the dispersion coefficient of bubble motion are strongly dependent on hole or nozzle diameter. The value of dispersion coefficient of bubble motion (D_b) increases with the hole diameter $(D_b \ [cm^2/s] = 170 \ d_h \ [mm])$ at constant column diameter (0.04 m) in case of batch liquid gas upflow bubble column (Ohki and Inoue, 1970). This is because the larger holes produce the larger bubbles which rise with higher velocity and disperse the liquid around them more strongly.
- The value of characteristic factor of velocity distribution parabolically decreases with the hole diameter $(k \propto 1/d_h$, where the proportionality constant is 11.76 for reactor diameter 40 mm). He concluded that this may be due to an inhomogeneous distribution of the bubble size, which is formed by increasing the hole diameter. In gas–liquid jet reactors with two-phase nozzles such as ejector, both gas and liquid are concentrically introduced and the kinetic energy of the liquid jet is utilized to disperse gas flow into fine bubbles. In the downflow system, the turbulence of the fluid inside the column is

much significant due to higher momentum transfer of the plunging liquid jet and higher gas hold-up (Majumder et al., 2005).

- As the nozzle diameter increased, the characteristic factor of velocity distribution follows parabolic pattern ($k \propto 1/d_n^{0.68}$), which is similar to upflow system (Ogawa et al., 1982) but the order is different.

- Dispersion coefficient of bubble motion in downflow system is increased with an increase in the nozzle diameter which follows the same trend as upflow system ($k \propto 1/d_n^{1.792}$). As the nozzle diameter increases, the dispersion coefficient of bubble motion also increases ($D_b \propto d_n^{1.68}$) due to the variation of velocity pattern with the increase in nozzle diameter.

- The order of k is much higher in upflow than downflow due to the intensity of turbulence of phases. They reported that more turbulence follows less k. In case of downflow system, due to the opposite force, the resistance of bubble buoyancy produces more turbulence with higher momentum exchange of fluid by liquid jet.

- The column diameter has the most significant influence on the fluid dispersion. The scale of liquid circulation is proportional to the column diameter and the dispersion coefficient is strongly influenced by the column diameter (Majumder, 2008). In case of upflow bubble column reactor, liquid is flowing upward in the centre and downward near the wall region. Obviously, due to turbulence, the up- and downflowing liquid interacts and will, therefore, exchange the liquid. This contribution is represented by the characteristic factor of the velocity distribution. As the column diameter increases, the interaction and formation of large circulation will increase, which may cause the decrease in velocity characteristic factor. The decrease in characteristic factor will increase the representative velocity which caused the increase in circulation with an increase in column diameter.

- The effect of liquid circulation enhances the dispersion coefficient of bubble motion with an increase in column diameter. The characteristic factor of velocity distribution (K) increases with the column diameter. This may be due to the mode of distribution of gas phase which is an important criterion to influence the velocity pattern in the column. At low gas flow rate, the larger diameter decreases the fluid circulation and turbulence and hence it increases the velocity characteristic factor. In the batch gas-aided slurry reactor, with low gas flow rate, it is maximum for the particular unit. This is due to the zero liquid velocity. As the column diameter increased, the liquid velocity in the column tends to zero, which turns the characteristic factor to constant and maximum. In case of downflow system (Majumder et al., 2005), the absence of liquid circulation with respect to column diameter results in the opposite effect on the characteristic factor of velocity and dispersion coefficient of bubble motion.

- The rate of liquid circulation increases with increasing superficial liquid velocity, which causes the increase in dispersion coefficient of bubble motion.

- The velocity characteristic factor depends on the intensity of turbulence.

- Superficial liquid velocity has an immense effect on the downflow system because as the liquid velocity increases the transfer of momentum by liquid jet in downflow system (Majumder et al., 2005) makes more turbulence and intense dispersion of phases, which resulted in deviation of D_b and K from upflow system.

- The intensity of liquid-phase dispersion increases with 1.2th power of superficial gas velocity (Ohki and Inoue, 1970). On the basis of energy balance and the assumption that a multiple number of circulation cells in axial direction occur and the dimensions are proportional to the column diameter.

- The gas hold-up is an important factor to affect the dispersion coefficient of bubble motion and velocity pattern. The dispersion coefficient of the bubble motion increases with an increase in gas hold-up and velocity characteristic factor decreases with an increase in gas velocity due to an increase in turbulence.

- For slow gas velocity, less gas hold-up decreases the turbulence of phase in a particular liquid velocity for constant geometric variables which sometimes increases the velocity characteristic factor. It has been observed that, in the downflow bubble column, the values of k are much less than that of upflow slurry bubble column.

- The dispersion process in the downflow slurry bubble column is also a consequence of large bubbles whose terminal velocities exceed the liquid velocity in the column. The terminal rise velocity of the bubble is governed by the bubble size. This may cause the greater effect of superficial gas velocity on the dispersion coefficient of bubble motion and coefficient of the velocity profile.

- The superficial gas velocity is directly proportional to superficial liquid velocity. Therefore, the trend for variation of D_b and K with superficial liquid velocity will be same as superficial gas velocity.
- The velocity characteristic factor is greater than $4Pe^2$. For the real value of the characteristic factor, K of velocity distribution will be greater than the square of double of the Peclet number ($[d_c u_o]$ / D_{sl}). If characterization factor, k, is equal to the square of double of the Peclet number, the coefficient of bubble motion is equal to half of the dispersion coefficient of liquid. An increase of dispersion coefficient of bubble motion, while decreases of characteristic factor ensure that the parameter $\sqrt{1 - \dfrac{4}{K}\left(\dfrac{d_c u_0}{D_{sl}}\right)^2}$ will be in the range of 0–1.

6.6   EFFICIENCY OF THE DISPERSION

Plunging jet with ejector system provides a high momentum exchange to result in high slurry dispersion and efficient gas dispersion. The gas dispersion and energy efficiency by plunging jets have often happened in practice (Bin, 1993; Majumder et al., 2005; Sivaiah and Majumder, 2012, 2013; Sivaiah et al., 2012). The gas dispersion is desirable to achieve gas absorption coupled with good dispersion in some multiphase reactors (Kara et al., 1982). In particular, due to encouraging energy requests, jet aerators have the potential application in many chemical, fermentation and waste water treatment processes as discussed in Chapter 1. Some of the practical applications of slurry bubble column with ejector system are coal utilization and conversion processes, such as direct coal hydroliquefaction and oxydesulphurization of coal, hydrogenation reactions (e.g. saturation of fatty acids), catalytic hydrogenation processes, hydrocracking, desulphurization of petroleum fractions, Fischer–Tropsch synthesis and methanol synthesis (Costa et al., 1986; Li et al., 2003). The plunging liquid jets also have an interesting application in the process of flotation of minerals. The interest and the scope of research on plunging liquid jet systems have increased considerably (Bin, 1993). The plunging jet reactors with downcomer are finding increasing importance due to their simplicity of scale-up, getting higher selectivity of the desired products and the maximum conversion of the reactants. However, the three-phase dispersion in the vicinity of the plunging jet and its efficiency to incorporate the gas entrainment efficiency based on the dispersion has not enunciated. In this chapter, the efficiency of the energy dissipation on the three-phase dispersion in the gas-interacting downflow slurry reactor is described. The energy efficiency of the gas dispersion device is calculated by the rate of kinetic energy supplied by the solid–liquid slurry jet divided by the volumetric gas dispersion rate (Bin and Smith, 1982). The rate of kinetic energy of the solid–liquid slurry jet (E_s) supplied can be calculated as

$$E_s = \frac{1}{2} \dot{m}_{sl} u_j^2 \tag{6.17}$$

where $\dot{m}_{sl}$ is the mass flow rate of solid–liquid slurry. As a function of volumetric flow rate of the slurry, Equation 6.17 can be written as

$$E_s = \frac{1}{2} Q_{sl} \rho_{sl} u_j^2 = \frac{1}{2} \left(\frac{\pi}{4} d_n^2 \right) u_j \rho_{sl} u_j^2 \tag{6.18}$$

On simplification of Equation 6.18, one gets

$$E_s = 0.125 \pi \rho_{sl} d_n^2 u_j^3 \tag{6.19}$$

The energy utilized for the gas–liquid–solid three-phase dispersion is equal to the product of energy supplied to the system and the energy utilization efficiency of dispersion. For the dispersion of gas in the slurry phase in dispersion zone, the energy utilization efficiency is equal to the ratio of the energy utilized for dispersion of the gas–liquid–solid mixture and the total energy supplied. Therefore, it can be expressed as

$$\eta_m = \frac{\text{Energy Uti lized}}{\text{Total Energy Supplied}}$$

$$= \frac{\text{Total Energy Supplied} - \text{Energy loss due to friction}}{\text{Total Energy Supplied}}$$

$$= \frac{\rho_{sl} u_j^2 / 2 - C_m \rho_{sl} u_j^2 / 2}{\rho_{sl} u_j^2 / 2}$$

$$= 1 - C_m \tag{6.20}$$

In the present downflow slurry bubble column, the gas–liquid–solid dispersion occurred when a slurry jet is discharged from the nozzle into the pool of slurry in the column. The energy balance in the dispersion zone can be written as

$$(\dot{m}_{sl} + \dot{m}_g)e_m$$
$$= (p_{atm})(Q_{sl} + Q_g) + [\dot{m}_{sl}(u_j^2 - u_m^2)/2] + [\dot{m}_g(u_{gd}^2 - u_m^2)] + (Q_{sl} + Q_g)\rho_m L_m g \qquad (6.21)$$

The energy dissipation rate of dispersion per unit volume in the gas–liquid–solid downflow system can be evaluated by Equation 6.22, developed for the plunging liquid jet bubble column (Evans et al., 1996)

$$e_m = \frac{u_j^2}{2}\left[1 - 2b - b^2(1+\lambda)^2 + 2b^2(1+\lambda)\right] \qquad (6.22)$$

where e_m is the energy dissipation rate, b is the area ratio of jet to column and λ is the gas-to-liquid volumetric flow ratio. To account for the energy loss in the dispersion zone, defining the energy dissipation coefficient of dispersion (C_m) as

$$\rho_{sl}e_m = C_m(u_j^2\rho_{sl}/2) \qquad (6.23)$$

By substituting Equation 6.23 into Equation 6.21 and simplification, one gets

$$C_m = (2/u_j^2\rho_{sl}(1+M_r))(1+Q_r)[p_{atm} + \rho_m L_m g] \\ - (1/A_r^2(1+M_r))(A_r^2 - (1+Q_r)^2 - M_r(2Q_r+1) \qquad (6.24)$$

where M_r is the mass ratio of gas to slurry, A_r is the area ratio of column to nozzle, Q_r is the volumetric flow ratio of gas to slurry and C_m is the fraction of jet energy dissipated. In the gas-interacting downflow of gas–liquid–solid three-phase flow, gas is sucked through the secondary entrance of the ejector by the high-velocity slurry jet as described in Chapter 1. The variation of gas dispersion rate (Q_g) with slurry concentration (w) at different rate of energy supplied (E_s) is shown in Figure 6.5. The gas dispersion rate decreases with an increase in the slurry concentration and increases with an increase in the rate of energy supply. This is due to the reason that on increasing the slurry concentration, the turbulence decreases in the column because of the increase in the apparent viscosity of the slurry. The viscous effect of the slurry of different concentration increases the pool resistance of the slurry inside the column, therefore the gas bubbles face more resistance to move in downward. Therefore, to obtain the same rate of gas dispersion with an increase in slurry concentration, higher energy is required. The energy requirement with slurry concentration varies approximately as $E_s \infty\ w^{1.877}$ (Sivaiah et al., 2012).

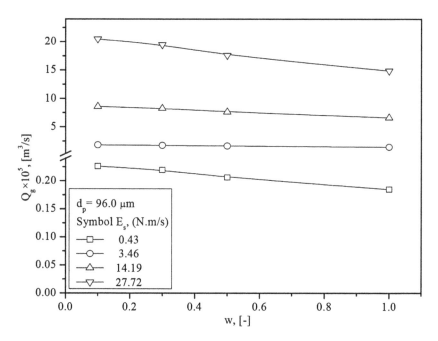

FIGURE 6.5   Variation of gas dispersion with slurry concentration (w).

The variation of energy utilization efficiency (η_m) for the dispersion of gas–liquid–solid mixture with slurry Reynolds number (Re_{in}) at different gas Reynolds numbers (Re_g) is shown in Figure 6.6. The energy utilization for gas–liquid–solid dispersion in the dispersion zone is increased with an increase in the slurry flow rate and decreased with an increase in the separator outlet gas flow rate. As the slurry flow rate increases, the kinetic energy of the liquid jet entrains more gas into the slurry, which results in an increase of turbulence of the gas–liquid–solid mixture in the column. As the turbulence in the column increases, the jet energy utilization is higher for the dispersion of gas–liquid–solid mixture in the dispersion zone because of an increase in the gas hold-up. At a constant slurry flow rate, as the superficial gas velocity increases from the separator outlet, the flowing gas bubbles move at a faster rate in a downward direction because of their less buoyancy effect. Consequently, the energy utilization for the dispersion of gas–liquid–solid mixture in the dispersion zone decreases due to lower turbulence experienced by the gas bubbles.

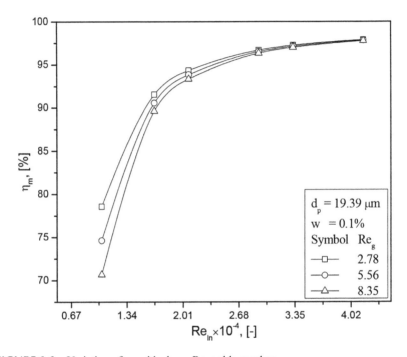

FIGURE 6.6   Variation of η_m with slurry Reynolds number.

The gas dispersion efficiency of the different downflow slurry bubble columns can be evaluated based on the knowledge of energy dissipation and gas hold-up in the column. The energy dissipation in the three-phase gas–liquid–solid downflow occurs because of the frictional losses between the phases in the column. The energy dissipation coefficient of dispersion (C_m) in the dispersion zone can be represented as a function of different dimensionless groups by developing a correlation based on experimental data as:

$$C_m = 5.034 \times 10^3 \ Re_n^{0.069} \ Re_g^{0.138} \ We^{-0.924} \ D_r^{-0.004} \qquad (6.25)$$

where Re_n is the Reynolds number based on nozzle diameter, Re_g is the gas Reynolds number, We is the Weber number and D_r is the diameter ratio of nozzle to particle. The ranges of the variables of Equation 6.25 are 9.55 × 103 < Re_n < 49.03 × 103, 2.78<Re_g < 8.35, 0.84 × 103 < We < 13.49 × 103 and 0.52 × 102<D_r<22.68 × 102. The correlation coefficient and overall standard error of Equation 6.25 are found to be 0.991 and 0.086, respectively. The parity of experimental and calculated values of energy dissipation coefficient of dispersion (C_m) is shown in Figure 6.7. Also, the variation of energy dissipation coefficient of dispersion with slurry Reynolds number at different gas Reynolds number is shown in Figure 7.3.

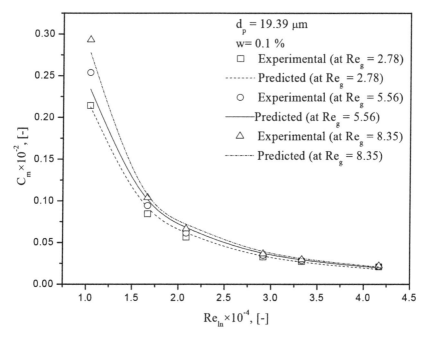

FIGURE 6.7 Variation of energy dissipation coefficient of dispersion with slurry Reynolds number.

With an increase in the slurry velocity, the intensity of the momentum of gas–liquid–solid mixture in the column increases and utilizes more energy for the dispersion of the gas–liquid–solid mixture in the column and then reduces the energy dissipation in the dispersion zone. The momentum of gas–liquid–solid mixture depends on the fluid circulation in the column. In the downflow system, liquid is moved in the downward direction, whereas gas is moved in both upward and downward directions as a dispersed phase of the bubble. The bigger bubbles, whose buoyancy force is greater than the downward momentum of the fluid, move upward, whereas the smaller bubbles move in a downward direction due to their smaller buoyancy. This phenomenon results in internal fluid circulation on the cell of fluid in the column. The fluid circulation cell increases with the column diameter. This fluid circulation results in the dispersion of fluid in the column. The intensity of dispersion can be interpreted by the Peclet number. The intensity of fluid circulation in the column can be calculated from the dispersion number (inverse of Peclet number), which can be estimated by RTD method (Sivaiah and Majumder, 2013). A general correlation for solid–liquid circulation velocity for the gas-interacting downflow slurry bubble column as a function of different dimensionless groups can be developed and represented by

$$u_c/u_{sl} = 1.683 \times 10^{-3} \, Re_{sl}^{1.321} \, Fr_g^{0.505} \qquad (6.26)$$

where Re_{sl} is the slurry Reynolds number $(u_{sl} d_c \rho_{sl}/\mu_{sl})$ and Fr_g is the Froude number $(u_g^2/g d_c)$.

KEYWORDS

- gas–liquid–solid mixture
- phase dispersion
- residence time distribution (RTD)
- Reynolds numbers
- velocity distribution model

REFERENCES

Aris, R. On the Dispersion of a Solute in a Fluid Through a Tube. *Proc. Roy. Soc. A London* **1956,** *235,* 67–77.

Bin, A. K. Gas Dispersion by Plunging Liquid Jets. *Chem. Eng. Sci.* **1993**, *48*, 3585–3630.

Bin, A. K.; Smith, J. M. Mass Transfer in a Plunging Liquid Jet Absorber. *Chem. Eng. Commun.* **1982**, *15*, 367–383.

Cao, C.; Dong, S.; Geng, Q.; Guo, Q. Hydrodynamics and Axial Dispersion in a Gas-Liquid-Solid EL-ALR with Different Sparger Designs. *Ind. Eng. Chem. Proc. Des. Dev.* **2008**, *47*, 4008–4017.

Costa, E.; Lucas, A. D.; Garcia, P. Fluid Dynamics of Gas-Liquid-Solid Fluidized Beds. *Ind. Eng. Chem. Process Des. Dev.* **1986**, *25*, 849–854.

Chen, P.; Gupta, P.; Dudukovic, M. P.; Toselandc, B. A. Hydrodynamics of Slurry Bubble Column During Dimethyl Ether (DME) Synthesis: Gas-Liquid Recirculation Model and Radioactive Tracer Studies. *Chem. Eng. Sci.* **2006**, *61*, 6553–6570.

Degaleesan, S.; Dudukovic, M. P. Liquid Backmixing in Bubble Columns and the Axial Dispersion Model. *AICHE J.* **1998**, *44*(11), 2369–2378.

Evans, G. M.; Jameson, G. J.; Rielly, C. D.; Free Jet Expansion and Gas Dispersion Characteristics of a Plunging Liquid Jet. *Exp. Therm. Fluid Sci.* **1996**, *12*, 142–149.

Gupta, P.; Ong, B.; Al-Dahhan, M. H.; Dudukovic, M. P.; Toseland, B. A. Hydrodynamics of Churn Turbulent Bubble Columns: Gas-Liquid Recirculation and Mechanistic Modeling. *Catal. Today* **2001**, *64*, 253–269.

Kang, Y.; Kim, S. D. Radial Dispersion Characteristics of Two and Three Phase Fluidized Beds. *Ind. Eng. Chem. Proc. DD* **1986**, *25*, 717–722.

Kara, S.; Kelkar, B. G.; Shah, Y. T.; Carr, N. L. Hydrodynamics and Axial Mixing in a Three-Phase Bubble Column. *Ind. Eng. Chem. Process Des. Dev.* **1982**, *21*, 584–594.

Kato, Y.; Nishiwaki, A. Longitudinal Dispersion Coefficient of a Liquid in a Bubble Column. *Int. Chem. Eng.* **1972**, *12*, 182–287.

Kato, Y.; Morooka, S.; Koyama, M.; Kago, T.; Yang, S. Longitudinal Dispersion Coefficient of Liquid in Three-Phase Fluidized Bed for Gas-Liquid-Solid Systems. *J. Chem. Eng. Jpn.* **1985**, *18*(4), 313–318.

Kim, S. D.; Kim, C. H. Axial Dispersion Characteristics of Three Phase Fluidized Beds. *J. Chem. Eng. Jpn.* **1983**, *16*, 172–178.

Kim, S. D.; Kim, H. S.; Han, J. H. Axial Dispersion Characteristics in Three-Phase Fluidized Beds. *Chem. Eng. Sci.* **1992**, *47*, 3419–3346.

Krishna, R.; Sie, S. T. Design and Scale-Up of the Fischer-Tropsch Bubble Column Slurry Reactor. *Fuel Proc. Technol.* **2000**, *64*(1–3), 73–105.

Lee, K. I.; Son, S. M.; Kim, U. Y.; Kang, Y.; Kang, S. H.; Kim, S. D.; Lee, J. K.; Seo, Y. C.; Kim, W. H. Particle Dispersion in Viscous Three-Phase Inverse Fluidized Beds. *Chem. Eng. Sci.* **2007**, *62*, 7060–7067.

Levenspiel, O. *Chemical Reaction Engineering,* 2nd Edn.; Wiley: New York, 1972.

Li, H.; Prakash, A.; Margaritis, A.; Bergougnou, M. A. Effect of Micro-Sized Particles on Hydrodynamics and Local Heat Transfer in Slurry Bubble Column. *Powder Technol.* **2003**, *133*, 171–184.

Lim, H. O.; Seo, M. J.; Kang, Y.; Jun, K. W. Particle Fluctuations and Dispersion in Three-Phase fluidized Beds with Viscous and Low Surface Tension Media. *Chem. Eng. Sci.* **2011**, *66*, 3234–3242.

Lu, W. J.; Hwang, S. J.; Chang, C. M. Liquid-Dispersion in Two- and Three-Phase Airlift Reactors. *Chem. Eng. Sci.* **1994**, *49*, 1465–1468.

Majumder, S. K. Analysis of Dispersion Coefficient of Bubble Motion and Velocity Characteristic Factor in Down and Upflow Bubble Column Reactor. *Chem. Eng. Sci.* **2008**, *63*, 3160–3170.

Majumder, S. K.; Kundu, G.; Mukherjee, D. Mixing Mechanism in a Modified Co-Current Downflow Bubble Column. *Chem. Eng.* **2005,** *112*, 45–55.

Manish, P.; Majumder, S. K. Quality of Mixing in Downflow Bubble Column Based on Information Entropy Theory. *Chem. Eng. Sci.* **2009**, *64*(8), 1798–1805.

Muroyama, S.; Hashimoto, K.; Kawabata, T.; Shiota, M. Axial Liquid Dispersion in Three Phase Fluidized Beds. *Kagaku Kogaku Ronbunshu* **1978**, *4*, 622–627.

Ogawa, S.; Kobayashi. M.; Tone. S.; Otake. T. Liquid Phase Mixing in the Gas-Liquid Jet Reactor with Liquid Jet Ejector. *J. Chem. Eng. Jpn.* **1982**, *15*(6), 469–473.

Ohki, Y.; Inoue, H. Longitudinal Mixing of the Liquid Phase in Bubble Columns. *Chem. Eng. Sci.* **1970,** *25*, 1–16.

Rubio, F. C.; Miron, A. S.; Garcia, M. C. C.; Camacho, F. G.; Grima, E. M.; Chisti, Y. Dispersion in Bubble Columns: A New Approach for Characterizing Dispersion Coefficients. *Chem. Eng. Sci.* **2004**, *59*, 4369–4376.

Saxena, S. C.; Rosen, M.; Smith, D. N.; Reuther, J. K. Mathematical Modeling of Fischer-Tropsch Slurry Bubble Column Reactors. *Chem. Eng. Commun.* **1986**, *40*, 97–151.

Sivaiah, M.; Majumder, S. K. Gas Holdup and Frictional Pressure Drop of Gas-Liquid-Solid Flow in a Modified Slurry Bubble Column. *Int. J. Chem. React. Eng.* **2012**, *10*(1), 1542–6580.

Sivaiah, M.; Majumder, S. K. Dispersion Characteristics of Liquid in a Modified Gas-Liquid-Solid Three-Phase Down Flow Bubble Column. *Particul. Sci. Technol.* **2013**, *31*, 210–220.

Sivaiah, M.; Parmar, R.; Majumder, S. K. Gas Entrainment and Holdup Characteristics in a Modified Gas-Liquid-Solid Down Flow Three-Phase Contactor. *Powder Technol.* **2012**, *217*, 451–461.

Tang, W. T.; Fan, L. S. Axial Liquid Dispersion in Liquid-Solid and Gas-Liquid-Solid Fluidized Beds Containing Low Density Particles. *Chem. Eng. Sci.* **1990,** *45*, 543–551.

Taylor, G. I. Dispersion of Soluble Matter in Solvent Flowing Slowly Through a Tube. *Proc. Roy. Soc. A London* **1953**, *219*, 186–203.

Turner, J. R.; Mills, P. L. Comparison of Axial Dispersion and Mixing Cell Models for Design and Simulation of Fischer-Tropsch Slurry Bubble Column Reactors. *Chem. Eng. Sci.* **1990,** *45*, 2317–2324.

CHAPTER 7

MASS TRANSFER PHENOMENA

CONTENTS

7.1 INTRODUCTION

Mass transfer by bubble dispersion in a slurry bubble column is widely occurred in biological, chemical and environmental applications. There is a great value in trying to intensify mass transfer with minimal energy consumption. Nowadays, bubble column with plunging jets are frequently met in practice as it provides high momentum exchange resulting in high-phase mixing and efficient gas dispersion. Some other advantages of these plunging jet bubble columns are: high dispersion efficiency and conversion per pass, produce uniform and finer bubbles, higher residence time of gas bubbles, reasonable interphase mass transfer rates at low energy input (Bin and Smith, 1982; Bin, 1993; Sivaiah and Majumder, 2013). Plunging liquid jet (with ejector system) is used to suck gas and distribute it by utilizing the jet kinetic energy. The distributed gas, as a dispersed phase of fine bubbles in the column, creates turbulence due to momentum exchange between phases. Moreover, due to counteract of the downward fluid momentum with the buoyancy of the gas bubbles, the bubbles reside for longer time in the reactor. Hence, higher interfacial area and residence time can be achieved in the plunging jet bubble column reactors which are attracting the investigators in recent time. Occurrence of gas–liquid or gas–liquid–solid or gas–solid multiphase flow is quite common in chemical process industries. Therefore, several investigators have selected the gas–liquid–solid flow phenomena over the last few years. Due to higher interfacial area, the plunging system is gaining importance for the industrial applications related to mass transfer operations (Patil et al., 1984; Yasunishi et al., 1988; Zaki et al., 1997; Mandal, 2010; Verma and Rai, 2003; Ramesh et al., 2010; Sivaiah and Majumder, 2013; Jin et al., 2014). The mass transfer operations in a multiphase system depend on the flow behaviour and the dispersion phenomena. In bubble column reactors, mass transfer phenomena are generally enunciated by phase resistance theory where solute is transferred from gas bubble to a continuous liquid phase. The intensity of mixing, bubble size distribution, and physical properties of the system are the influencing factors for the mass transfer in gas-aided slurry reactors. Based on these factors, different investigators developed different correlations to interpret the mass transfer coefficient in bubble column reactor (Yang et al., 2007; Sivaiah and Majumder, 2013). The degree of mass transfer is determined by the mass transfer coefficients.

Sharma and Mashelkar (1968) first reported increase of the gas absorption rate by small gas-absorbing particles in a bubble column. Lee and Tsao (1972) also have observed the same trend in a stirred slurry reactor. Alper et al. (1980) studied the mass transfer phenomena by active carbon particles, and concluded that the two-film model of Whitman (1923) and the penetration model do not fit well for their experimental observation of mass transfer phenomena in the presence of solid particle. In slurry reactors, particle-to-bubble adhesion (PBA) affects the gas hold-up, the bubble size distribution and the bubble coalescence rate. From the literature, it is seen that all the present relevant work published by different authors are based on two-phase systems in upflow column. Studies on mass transfer characteristics in a gas-interacting liquid–solid co-current downflow bubble column reactor are scanty.

7.2 MECHANISMS OF MASS TRANSFER IN A GAS-INTERACTING SLURRY REACTOR

From the literature, it is seen that the enhancement if degree of gas–liquid mass transfer in presence of particles suspended in the liquid phase has been investigated by several authors. Several models/mechanisms have been proposed to interpret the enhancement of mass transfer (Alper et al., 1980; Holstvoogd et al., 1988; Lindner et al., 1988; Wimmers et al., 1988; Vinke et al., 1991; Beenackers and van Swaaij, 1993; Tinge and Drinkenburg, 1995; van der Zon et al., 1999; Ruthiya et al., 2005; Mandal, 2010; Sivaiah and Majumder, 2013; Meng et al., 2017). An important aspect of gas–liquid mass transfer enhancement appears to be the sticking of particles to the gas–liquid interface (Wimmers et al., 1988; Vinke et al., 1991, 1993). The attachment is termed as particle to bubble attachment as shown in Figure 7.1 schematically.

This attachment depends on the liquid properties, surface-active components and nature of liquid either aqueous or organic, particle characteristics, adsorption capacity between liquid and process parameters (e.g. turbulence intensity, particle concentration). Ruthiya et al. (2003) explicitly reported the mechanism of gas–liquid mass transfer enhancement in the presence of solid particle which can be summarized as follows:

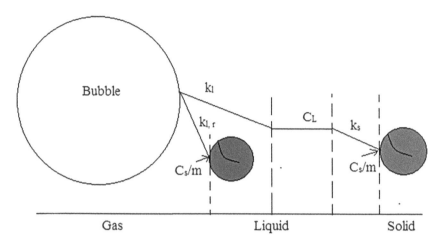

FIGURE 7.1	Schematic representation of mass transfer in gas–slurry system.

7.2.1 BOUNDARY LAYER MIXING

Four different physical phenomena influence the convective mass transfer and the concentration gradient at the gas–liquid interface (Ruthiya et al., 2003): (i) the effective gas–liquid boundary layer thickness is reduced by collisions of the particles with the boundary layer increasing the gas–liquid mass transfer coefficient (k_l); (ii) a local degree of turbulence at the gas–liquid interface is induced by large particles (if particle diameter is greater than gas–liquid film thickness at liquid side of the interface) to increase the refreshment rate of the liquid in the gas–liquid boundary layer by mixing with the bulk liquid. However, particles may reduce the turbulence at the gas–liquid interface, leading to a decrease of liquid side mass transfer coefficient; (iii) the mass transfer coefficient depends on the rate of coalescence of gas bubbles due to the fact that bubble coalescence by particles is especially induced at low mixing intensities, and it may give rise to larger bubbles that have a more mobile interface and result in larger mass transfer coefficient; bubble coalescence causes re-dispersion of entrained gas, which means additional surface renewal and more mass transfer; and (iv) small particles (if particle diameter is less than gas–liquid film thickness at liquid side of interface) may decrease mass transfer due to a decrease in the actual liquid hold-up for diffusion of gas at the interface. The number of particles adhering to bubbles is dependent on the balance between shear stress and attachment forces. The shear stress is proportional to the superficial gas velocity and the kinetic energy utilized to disperse the bubble in a bubble column. When the shear forces become higher than the attachment induced forces, particles are removed from the gas–liquid interface. The relative effect of particle to bubble attachment on the increase of the rate of mass transfer will, therefore, decrease at higher mixing intensities.

7.2.2   SHUTTLING

The penetration of liquid film is one of the important characteristic factors for adsorption of the gas onto the particle surface. Particles with a high specific surface area and porosity penetrate the liquid film at the gas–liquid interface and may adsorb some of the dissolved gas (Ruthiya et al., 2003). The adsorbed gas will be desorbed after transporting the particles to the bulk liquid. The transport of the gas to the bulk liquid is increased by intense mixing of the phases. In gas-aided slurry reactors, an increase of the volumetric gas–liquid mass transfer coefficient ($k_l a_l$) is regarded to an increase of mass transfer coefficient of about 200–300% (Kars et al., 1979; Alper et al., 1980), whereas Quicker et al. (1987) reported that it is increased from 20 to 50% due to this shuttle effect. This mechanism is dependent on the partition coefficient (m) of the particles and on the residence time of the particles in the gas–liquid boundary layer. If a significant amount of particles is present at the interface, mechanism 2 prevails in case of particles of size equal or smaller than the gas–liquid boundary layer (Ruthiya et al., 2003; Beenackers and van Swaaij, 1993; Schumpe et al., 1987). They reported that increasing the particle concentration and with increasing mixing intensity, the visiting frequency of particles at the gas–liquid interface will increase, leading to an increased transport of gas from the gas–liquid interface to the bulk liquid, which will result in a larger value of mass transfer coefficient.

7.2.3   COALESCENCE INHIBITION

Particles adhering to gas bubbles in the slurry can reduce or hinder coalescence of gas bubbles. This increases the value of the gas–liquid interfacial area, hence mass transfer. Different variables that influence this type of mass transfer are: (i) surface tension of liquid; (ii) viscosity and density of liquid/slurry; (iii) ionic forces; (iv) lyophobicity of particles (wetta-bility) and (v) particle size. Lindner et al. reported that a gas hold-up can be increased by an electrolyte (salt solution) due to decrease of bubble coalescence rate with electrolyte. With increasing superficial gas velocity or stirrer speed, the shear stresses in the system increase, decreasing the effect of electrolyte and solid particles on the increase of the gas–liquid interfacial area (Ruthiya et al., 2003). Ionic forces or the local electrostatic potential at the gas–liquid interface slows down the film drainage speed between two bubbles which result in a lower rate of bubble coalescence and an increased number of smaller bubbles (Marrucci, 1969; Prince and Blanch, 1990; Ruthiya et al., 2003). According to Schumpe et al. (1987), small carbon particles have a coalescence hindering effect on very small 'ionic' bubble clouds. Jamialahmadi and Muller-Steinhagen (1991) reported that wettable particles tend to repel the gas interface, therefore acting as a buffer between two adjacent gas bubbles. This stabilizes small bubbles and therefore the formation of large bubbles by coalescence is decreased. The opposite effect is also reported for non-wettable particles (using polypropylene). In this regard, the results of Quicker et al. (1987) are contradictory and explained the increase in k_l to the shuttling of particles.

7.2.4 BOUNDARY LAYER REACTION OR GRAZING EFFECT

In this case, when a chemical reaction occurs at the gas–liquid interface in presence of small catalyst particles, significant conversion occurs within the diffusion layer around the gas bubbles, thereby increasing the rate of mass transfer. As the concentration of gaseous reactants in the film layer is higher than in the bulk liquid, the reaction rate in the film layer will be higher (Ruthiya et al., 2003). Mass transfer enhancement during a reaction is a function of the lyophobicity and activity of the catalyst particles and of the turbulence intensity in the reactor (Lindner et al., 1988; van der Zon et al., 1999; Ruthiya et al., 2003). Though much research has already been documented on gas–liquid mass transfer enhancement, knowledge of the exact mechanism of increase in the gas–liquid mass transfer coefficient, the gas–liquid interfacial area or enhancement due to chemical reaction is still scanty. The studies on mass transfer enhancement for hydrogen gas in organic liquids, hydrogenation reactions in slurry systems and many more in the gas-interacting downflow slurry reactor can be done.

7.3   MEASUREMENT OF MASS TRANSFER COEFFICIENT

The product of mass transfer coefficient k_L and specific interfacial area a (based on dispersion volume) is known as volumetric mass transfer coefficient ($k_L a$). Mass transfer characteristic is one of the important parameters to design a reactor. When the gas side mass transfer resistance ($1/k_G H$) is insignificant then the mass transfer properties can be characterized by k_L alone. Bubble columns are widely used for absorption of gas into liquid for slow reactions and may be used for fast reactions if provided good heat removal facility (Deckwer, 1992). Volumetric mass transfer coefficient can be measured by physical method or chemical method.

7.3.1   PHYSICAL METHOD TO DETERMINE VOLUMETRIC MASS TRANSFER COEFFICIENT

Two physical methods are available in literature namely unsteady-state method and the stationary method. Both of these methods are based on some assumptions. They are as follows:

i)   Perfect mixing takes place in liquid phase.
ii)   There is a uniform driving potential for mass transfer in the column.
iii)   Saturated concentration of gas in liquid does not depend upon height (as equilibrium concentration depends on pressure).
iv)   There is no time delay in response of gas analyser to a change in dissolved gas.

7.3.1.1   UNSTEADY-STATE OR DYNAMIC METHOD

This method is used when purely physical mass transfer occurs and no chemical reaction takes place in the reactor. In most of the literature, oxygen is used as gas-phase fluid. Initially, the liquid is deoxygenated by passing nitrogen gas. Sodium sulphate with a cobalt chloride catalyst can also be used to deoxygenate the water (Chu et al., 2008). Then, the oxygen is passed in the liquid. The concentration of oxygen in the liquid can be measured by some digital dissolved oxygen meter or by Winkler's method (Kikuchi et al., 2009). In most of studies, the oxygen concentration is measured by inserting the polarograph oxygen electrode at the centre of dispersion height in the column. The mass transfer rate is, in this case, represented by

$$\frac{dC}{dt} = k_l a \left(C^* - C_l \right)$$ (7.1)

On integration of Equation 7.1 with boundary condition, at $t=0$, $C_l=C_0$ and at $t=t$, $C_l=C_t$, one can get

$$\ln \left(\frac{C^* - C_0}{C^* - C_t} \right) = k_l a t$$ (7.2)

Where C^* is the saturated concentration of gas in liquid, C_0 is initial concentration of gas in liquid, C_l is concentration of gas in liquid at time t. The term k_l is liquid-side mass transfer coefficient and a is specific interfacial area with respect to dispersed volume. Equation 7.1 holds when response time of measuring electrode, τ_p (time needed to record 63% of a stepwise change), is much less than the mass transfer response time ($1/k_l a$).

$$\text{That is } \tau_p << \frac{1}{k_l a}$$ (7.3)

Response times of electrode in between 0 and 5 s generally do not create much error and hence it is ignored.

7.3.1.2 STATIONARY METHOD

This method is also used when physical mass transfer occurs. In this method, liquid phase is passed through the reactor and the concentration of gas is measured at inlet and outlet of the reactor. If it is assumed that complete liquid mixing takes place and saturated concentration of gas in liquid does not depend upon height, then the volumetric mass transfer coefficient can be calculated as (Decker, 1992):

$$k_l a = \frac{u_{sl}}{L} \left(\frac{C_l - C_0}{C_l^* - C_l} \right)$$ (7.4)

where u_{sl} is superficial velocity of liquid in column, L is length of column reactor and C_0 and C_L are concentrations of gas in liquid at inlet and outlet of the reactor, respectively.

7.3.2 CHEMICAL METHOD FOR DETERMINATION OF VOLUMETRIC MASS TRANSFER COEFFICIENT

Generally, there are two chemical methods available in literature for the determination of volumetric mass transfer coefficient.

7.3.2.1 BY DANCKWERTS-PLOT

If a gas–liquid reaction takes place in the zone between diffusion and fast reaction regime, and the reaction rate can be varied by the fact such as change in catalyst concentration without changing the interfacial area to any considerable extent, then for a second order reaction with liquid-phase reactant (B), the rate of absorption is given as

$$R_A a = aC_A^* \left(D_A^* k_2 c_\kappa C_B^0 + k_l^2 \right)^{0.5}$$

(7.5)

where R_A is specific absorption rate or absorption rate per unit interfacial area (mol/m²s), $R_A a$ is rate of absorption or absorption rate per unit dispersion volume. C_A^*, k_2, c_κ, D_A^* and C_B^0 are gas-phase saturated concentration, reaction rate constant, catalyst concentration, diffusive coefficient of gas phase and concentration of liquid in bulk of gas phase, respectively. Plot of $(R_A a)^2$ versus $(k_2 c_\kappa C_B^0)^2$ gives a straight line with slope of $(aC_A^*)^2 D_A^*$ and an ordinate $(k_L a C_A^*)^2$. This is known as 'Danckwerts-plot' (Danckwerts, 1950). Both interfacial area and liquid side mass transfer coefficient can be calculated from the plot.

7.3.2.2 BY MEASURING ABSORPTION RATES

In this method, when an absorbed gas undergoes a chemical reaction of known kinetics, the volumetric mass transfer coefficient k is measured from the gas absorption rate. Suppose under some condition, the reaction between A and B occurs in the bulk and no reaction occurs in film, and also the reaction between A and B is sufficiently fast to ensure that the concentration of unreacted A in the bulk liquid is zero (Juvekar and Sharma, 1973) then,

$$A + B \rightarrow \text{Product} \tag{7.6}$$

With order m with respect to A and n with respect to B, the kinetics of reaction can be written as:

$$r_A = k_{m,n} C_A^m C_B^n \tag{7.7}$$

Then local absorption rate from the absorption-reaction theory can be given as:

$$R_A a = a C_A^* \left(\frac{2}{m+1} D_A^* k_{m,n} C_B^n C_A^{m-1} \right)^{0.5} \tag{7.8}$$

If the reduction of B is not much more at the interfacial area than in bulk, that is, the concentration of B at the interface is practically same as in bulk liquid then the reaction is of the pseudo-mth order:

$$k_m = k_{m,n} C_{B0}^n \tag{7.9}$$

where C_{B0} is concentration of B in bulk liquid phase. Now, by substituting the value of k_m in Equation 7.8, one can get

$$R_A a = k_l a C_A^* Ha \tag{7.10}$$

where the Hatta number (Ha) is defined as:
If $Ha \ll 1$ and $k_l a C_A^* \ll \varepsilon_l k_{m,n} C_A^* C_{B0}^n$, the rate of absorption is given by:

$$R_A a = k_l a C_A^* \tag{7.11}$$

If both the conditions are satisfied, then volumetric mass transfer coefficient can be calculated from absorption rate. Carbon dioxide absorption in caustic solution and sulphite oxidation catalysed by cobalt ion (Co^{2+}) are widely used for the determination of volumetric mass transfer coefficients by chemical method. Interfacial area is one of the important parameters as the rate of absorption is directly related to interfacial area. The reactor output is influenced by interfacial area. It is a function of reactor geometry, operating parameter and physical and chemical properties of both the phases. For slow and instantaneous absorption, mass transfer rate is determined by volumetric mass transfer coefficient $(k_L a)$ which is a function of specific interfacial area (a). The specific interfacial area (a) is the ratio of gas-phase component interfacial area (A_g) to dispersion volume (V_D) (Deckwer, 1992) which is represented by

$$a = \frac{A_g}{V_D} = \frac{A_g \varepsilon_g}{V_g} \tag{7.12}$$

where V_g is volume of gas phase and ε_g is gas hold-up. If there are n_b numbers of bubbles in the bubble diameter range of d_b, then the surface area of bubble formed within a specific dispersion volume is given as

$$A_g = \pi \sum n_b d_b^2 \tag{7.13}$$

The volume of gas phase formulated in dispersion given as

$$V_g = \left(\frac{\pi}{6}\right) \sum n_b d_b^3 \tag{7.14}$$

Now, by substituting the values of Equations 7.13 and 7.14 in Equation 7.12, one can get

$$a = \frac{6\varepsilon_g}{d_{bs}} \tag{7.15}$$

Specific interfacial area with respect to dispersion volume (a') is related with specific interfacial area with respect to liquid volume (a) and liquid hold-up (ε_l) as per Equation 7.15, dispersion volume can be calculated as

$$a' = \frac{a}{\varepsilon_l} \tag{7.16}$$

There are two methods to determine the interfacial area, namely physical and chemical method.

a) **Physical methods:** There are two physical methods which are commonly used to determine interfacial area. They are as follows:

i) **Indirect method**: If the gas hold-up and Sauter mean bubble diameter are known, then the specific interfacial area can be calculated by using Equation 7.15.

ii) **Direct methods**: The direct methods are optical in nature. Light transmission method is the optical method used to measure the interfacial area. This method was first used by Vermuelen et al. in 1955. The bubble size distribution does not affect the fraction of light transmitted by dispersion. The fraction of light transmitted (F_{lt}) is directly related to interfacial area and optical path length (L_p) (Sridhar and potter, 1978). The fraction of light transmitted (F_{lt}) is expressed by

$$F_{lt} = \exp\left(\frac{-aL_p}{4}\right) \tag{7.17}$$

Transmission method holds for $aL_p < 20$. Other direct methods involve refracted, diffracted or reflected measurement on bubble surface, whereby the reflected light flux or that penetrates through the liquid–gas dispersion is measured (Deckwer. 1992)

b) **Chemical method**: Physical process measures the local value of interfacial area, whereas by using chemical method, whole reactor volume can be evaluated. Generally, two chemical methods for determination of interfacial area are available in literature, first by Danckwerts-plot and the second by measuring absorption rates (Deckwer, 1992).

7.3.3 LIMITING CURRENT DENSITY METHOD

Several methods are available to estimate the mass transfer coefficient in a multiphase system of different geometries. The estimation of the mass transfer coefficient at vertical cylinders by limiting current density method is one of the easiest chemical methods (Ravoo et al., 1970). Many electrochemical processes such as electrowinning, electrodeposition, electroplating are carried out in electrochemical reactors in which a metal plate acts as an electrode and generally the container wall acts as another electrode. The chemical reactions take place on the surface of the electrode which is basically diffusion controlled (Sivaiah and Majumder, 2013). The study of these diffusion-controlled electrode reactions is essential in the design and scale up of such electrochemical reactors. Prior to the assembly of the test section, the surfaces of the point electrodes have to be polished and cleaned thoroughly. The electrical connectivity of the point electrode is to be checked using a digital meter. The current-potential measuring circuit is consisted of a voltmeter, an ammeter, rheostat, commutator, and a direct current source for connecting any desired electrode to the electrical circuit. The measurements are to be carried out for the case of an electrochemical redox system in the presence of an excess indifferent electrolyte (sodium hydroxide). This system has been chosen for these studies because (i) the chemical polarization involved is negligible, (ii) the reacting surface remains smooth and unaffected unlike in the cases of solids dissolution or sublimation processes and (iii) the measurements are relatively fast, accurate and reproducible. The measurement of limiting current is made in the lines similar to those reported earlier in the studies on ionic mass transfer (Lin et al., 1951). From the measured value of limiting current at any given electrode area, the mass transfer coefficients are evaluated. The detail of operational procedure is described by Sivaiah and Majumder (2013). In any electrochemical cell, the mass transfer between the bulk solution and the polarized surface takes place either by diffusion of ions due to concentration gradient or by migration of ions due to potential difference. By adding an excess indifferent electrolyte, the ionic migration of the reacting species can be eliminated and the electrode reaction is made diffusion controlled (Sivaiah and Majumder, 2013). The excess indifferent electrolyte can be chosen as 0.5 N of sodium hydroxide. The emphasis is, thus, laid on the diffusion of the reacting material towards electrode surface and the reacted species to the bulk solution. This process is of the type steady-state diffusion of solute in a dilute solution through a stagnant solvent (Sivaiah and Majumder, 2013). The transfer rate of diffusion of particular ionic species is written by using the Fick's first law of diffusion, which is written by the following expression:

$$\frac{ds}{dt} = \frac{A_e D_i (C_i - C_s)}{\delta (1 - t_i^+)} \tag{7.18}$$

where ds/dt is the transfer rate of ionic species (equivalents/s), A_e is the exposed area of the electrode surface, D_i is the diffusion coefficient of the ions, δ is the thickness of the hypothetical diffusion layer, C_i is the concentration of the ionic species on solution side of the diffusion layer, C_s is the concentration of the ionic species at electrode surface and t_i^+ is the transport number of the reacting species. If the concentration of indifferent electrolyte is very high compared to the concentration of the reacting ionic species, the transport number of the discharge ions can be taken as zero $t_i^+ = 0$). Therefore, Equation 7.18) can be expressed as

$$\frac{ds}{dt} = \frac{A_e D_l (C_i - C_s)}{\delta} \qquad (7.19)$$

The rate of discharge of ions can be expressed as:

$$\frac{i_d}{zF} = \frac{D_l (C_i - C_s)}{\delta} \qquad (7.20)$$

where i_d is the limiting current density (i/A_e), z is the number of electrons involved in the discharge process and F is the Faraday constant (96,500 C/ equivalent). The limiting current density represents the maximum rate at which the particular ion can be discharged under the given experimental conditions. At this current density, rapid increase of potential is observed (Sivaiah and Majumder, 2013). The value of the limiting current can be obtained from the plots drawn between current and applied potential. The attainment of the limiting current is indicated by a sharp increase in potential for a marginal increase in the current. Figure 7.2 illustrates the graphical method of obtaining limiting current from voltage versus current data.

When the limiting current is reached on a polarized electrode, the concentration of the ionic species at the electrode surface (C_s) becomes zero. Then Equation 7.20 can be written as

$$\frac{i_d}{zF} = \frac{D_l C_i}{\delta} \qquad (7.21)$$

The mass transfer coefficient (k_l) is defined as the ratio of diffusion coefficient to the thickness of the hypothetical diffusion layer (D_l/δ), which on substitution into Equation 7.21 yields

$$k_l = \frac{i_d}{zFC_i} \qquad (7.22)$$

The gas phase is used essentially to favour the hydrodynamic conditions at the reacting surface. A point electrode fixed at the inside wall of a column reactor is chosen as a reacting surface. The reacting ion in the continuous phase reaches the reacting surface under an electric potential by diffusion which depends on the polarity induced at the electrode surface. Reduction of ferricyanide ion was chosen for the studies which takes place at the electrode surface.

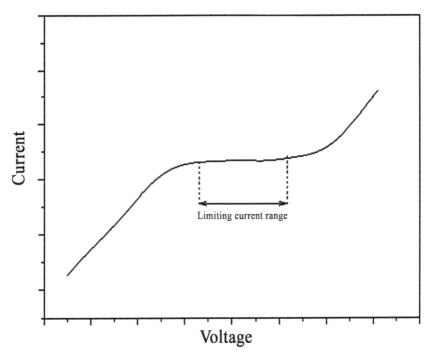

FIGURE 7.2   Graphical representation of limiting current identification.

7.4   EFFECT OF VARIABLES ON MASS TRANSFER COEFFICIENT

The mass transfer coefficient increases with increase in slurry velocity as well as gas velocity as shown in Figure 7.3. As the superficial slurry velocity increases, the jet momentum results the secondary air to be dispersed into discrete bubbles which, in turn, increase the turbulence and intensity of mixing in the column. This leads to further augmentation in the mass transfer rate. Moreover, at higher superficial liquid–solid slurry velocity, gas hold-up increases due to higher entrainment of the gas. As a result, the circulation and interaction of the gas and liquid phases increase inside the column and the flow gets more agitated. With an increase in superficial gas velocity, the coarser bubbles immediately get released from the jet due to their higher buoyancy, and these bubbles face more resistance to move in downward direction and spend more time in the column which results in more mixing due to the interaction of phases and, hence increase the mass transfer coefficient. The population of the bubble is higher due to the formation of more uniform finer bubbles with the interaction of the phases. The intimate contact between the phases increases as the gas hold-up increases with superficial gas velocity creating more turbulence in the column. This may result in increase in the mass transfer coefficient with increase in superficial gas velocity.

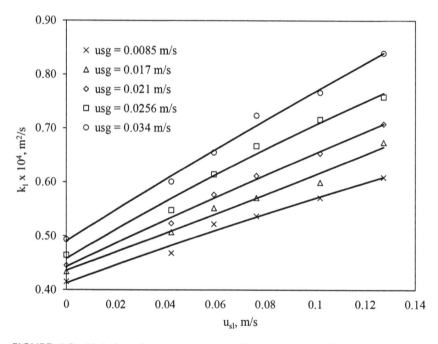

FIGURE 7.3   Variation of mass transfer coefficient with superficial slurry and gas velocities at particle diameter 19.39 μm and solid 0.1 wt.%.

The variation of mass transfer coefficient with slurry concentration and particle diameter is shown in Figure 7.4. The mass transfer coefficient is decreased with increase in slurry concentration and particle diameter. Increase in solids concentration increases the apparent viscosity of the liquid, thereby decreases the surface renewal rate and mobility of the particles and, hence mass transfer coefficient. Higher viscosities of the slurry also decrease the interfacial area due the coalescence of small bubbles which, in turn, decrease the mass transfer coefficient. Mass transfer coefficient values are higher for small size particles due to increase in liquid circulation with particle mobility and velocity inside the column. The mass transfer coefficients increased with increase in gas hold-up as same as those observed in case of gas velocity. The average gas hold-up increases almost linearly with increasing superficial gas and slurry velocity as described in the earlier chapter. As the gas hold-up increases, more bubbles occupy the column which decreases the flow area of the liquid and increases the true velocity of the liquid–solid slurry. Moreover, with increasing gas hold-up, the mean bubble diameter decreases and increases the specific interfacial area. Increase in interfacial area and true velocity of the slurry increase the turbulence in the column and decrease the thickness of the hydrodynamic boundary and diffusion layers. This leads to increase the electron migration towards the electrode surface and, hence increase the mass transfer rate. The change of surface properties of the liquid affects the bubble coalescence and generation and hence the interfacial area and mass transfer rate in the column.

The effect of a solid phase on gas–liquid mass transfer has been widely investigated, but this effect is very difficult to predict, especially for small particles at low concentrations, because mass transfer coefficient is found to depend on the size, density and shape of the particles, the wettability, the adsorption properties, and the solid concentration. All authors agree on a great decrease of volumetric mass transfer coefficient at high concentrations of solid particles (~10% wt.); this effect is often attributed to an increase of the apparent viscosity. For small concentrations of solid particles, the coefficient can be enhanced by solid particles by several phenomena like interactions between gas bubbles and large solid particles (Lee, 1973) using the Weber number (ratio of inertial force to interfacial force). If We > 3, the particle is able to divide or to stretch the bubble, generating a larger interfacial area. However, this relation uses the relative velocity between gas bubble and particle (Dietrich et al., 1992). Large particles (200–500 pm) can also create turbulence effects on the gas–liquid film whose thickness could diminish (Oguz et al., 1983). Very small particles (<0.4 μm) can rigidify the interface and behave like a soluble surfactant reducing the coalescence rate (Midoux and Charpentier, 1981, Chapman et al., 1983). The presence of small adsorbing particles in suspension can also increase the mass transfer as per shuttle mechanism (Alper et al, 1980;

Alper and Ozturk, 1986). In this case, the particles move from the inter-face to the liquid bulk and vice versa, the gas being adsorbed at the inter-face and desorbed in the liquid bulk. Dietrich et al. (1992) investigated the influence of solid particles on the volumetric mass transfer of hydrogen with two liquids, pure water and the hydrogenation mixture, and two kinds of solids, pyrophoric and non-pyrophoric Raney nickel particles. Pyroph-oric Raney nickel is loaded with adsorbed hydrogen and is very active for adsorption of other compounds. On the contrary, non-pyrophoric Raney nickel has been oxidized and is catalytically inert. They reported that the magnetism of pyrophoric Raney nickel causes the agglomeration of parti-cles as they proved by laser measurements in quiescent settling, pyroph-oric Raney nickel aggregates of about 250 μm are formed from 10–15 μm elementary particles (Dietrich et al., 1992).

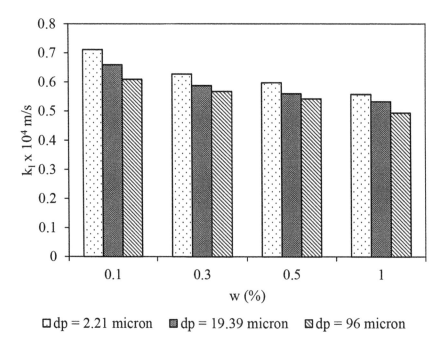

$\square\, dp = 2.21$ micron $\boxtimes\, dp = 19.39$ micron $\boxtimes\, dp = 96$ micron

FIGURE 7.4   Variation of mass transfer coefficient with slurry concentration and particle size at $u_{sg} = 0.0085$ m/s and $u_{sl} = 0.1019$ m/s.

The viscosity of the slurry varies with the slurry concentrations. The mass transfer coefficient is decreased with increase in slurry viscosity as similar to those in the case of slurry concentration. The variation of mass transfer coefficient with slurry viscosity at different superficial gas velocities is shown in Figure 7.5. Increase in slurry viscosity results in decrease the degree of turbulence in the liquid, thereby increases the thickness of the laminar film at the boundary layer between the electrode surface and the slurry (Sivaiah and Majumder, 2013). Increase in thickness of the laminar film decrease the electron deposition on the electrode surface which results in decreasing the mass transfer coefficient. The efficiency of bubble column based on mass transfer depends on operating pressure because the overall gas hold-up and the overall mass transfer coefficient have strong dependency on operating pressures irrespective of distribution of gas in the reactor (Majumder, 2008).

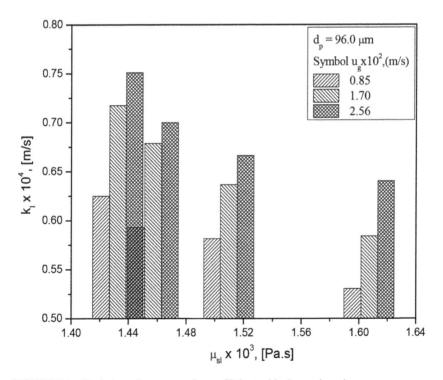

FIGURE 7.5 Variation of mass transfer coefficient with slurry viscosity.

The gas hold-up and volumetric mass transfer coefficient increased with the superficial gas velocity and the operating pressure. The volumetric mass transfer coefficient increased greatly due to the smaller bubble size generated at high pressures. The liquid side mass transfer coefficient, k_l, unlike $k_l a$, decreased with the increase of pressure, especially from ambient pressure to 0.4 MPa. The apparent decrease of k_l is attributed to the change in bubble dynamics and hydrodynamics (Majumder, 2008). At high pressure, smaller bubble sizes have lower slip velocity. Therefore, smaller bubbles have longer exposure time in the liquid interface renewal and cause decrease in k_l values at high pressure. The volumetric mass transfer coefficient increases with the superficial gas velocity, which, in turn, systematically induces an increase in specific interfacial area (Majumder, 2008). An increase in gas flow rate (superficial gas velocity) produces a clear increase on interfacial area due to the higher gas volume fed to contactor. At low liquid-phase flow rate, an increase in superficial gas velocity causes an increase in mass transfer coefficient due to a higher turbulence in the liquid phase and then a better surface renewal that increases mass transfer (Álvarez et al., 2008). At high values of liquid flow rate, the effect of superficial gas velocity is negligible because the influence of liquid flow rate is higher than that of the gas flow rate. An increase in liquid-phase flow rate produces a clear increase in the value of mass transfer coefficient due to the higher driving force existed in liquid phase but the ratio of $k_l a/u_{sl}$ decreases with increase in superficial liquid velocity which results in decrease in mass transfer efficiency in the bubble column reactor. The bubble size distribution has significant effects on mass transfer efficiency, because it determines the interfacial area per volume of the gas phase. The bubble size distribution is determined by the bubble coalescence and break-up. In a given system, bubble coalescence and break-up rates are mainly affected by the local gas hold-up and turbulent energy dissipation rate. Owing to the non-uniform radial profiles of the gas hold-up (Majumder, 2008) and dissipation rate, especially in the heterogeneous regime, the bubble size distribution varies not only with the superficial gas velocity but also with the radial position. According to Wang and Wang (2007), at low superficial gas velocities, the bubble size distributions are very similar at different radial positions. With an increase in the superficial gas velocity, the difference in the bubble size distributions at different radial positions becomes pronounced, especially after the flow enters the heterogeneous regime where bubbles are continuously breaking and coalescing and flowing with a mean stable bubble diameter. The interfacial area depends on the mean stable bubble diameter. Increase in interfacial area increases the efficiency of mass transfer efficiency. At high liquid velocity, in the plunging point of jet for gas formation, very large bubbles are formed and the small bubbles population fraction increases with an increase in height from the source of generation. These small bubbles probably result from the break-up of the large unstable bubbles. When the liquid velocity increases, the bubble diameter tends to decrease due to the breakage by high kinetic energy of the liquid jet. As gas-phase dispersion and the bubble size distribution, the liquid media have also a strong effect on mass transfer efficiency, but the liquid media effect seems more complex and is still disputed. In fact, the bubble size strongly depends on coalescence behaviour of the liquid, but the influence of the liquid properties on bubble coalescence and break-up remains difficult to quantify, especially in complex media. The most analysed liquid properties are viscosity and surface tension. A decrease in surface tension (due to surfactant addition) diminishes the bubble coalescence frequency: the bubbles are then smaller, slower and also more spherical. Generally, in the gas-interacting slurry reactor is referred to as heterogeneous flow reactor. In heterogeneous regime, the bubble size is governed by the coalescence–break-up equilibrium. The surface tension effect is similar to trend for single bubbles: a decrease in surface tension decreases bubble size and bubble velocity induces higher gas hold-up and higher mass transfer. The surface tension effect is particularly effective in homogeneous and transition regime and less in the heterogeneous regime where the reduction of coalescence is over-shadowed by the predominant effect of macro-scale turbulence. Moreover, with the increase in surface tension, the transition of bubbling regimes is delayed to higher gas velocity, whereas the heterogeneous regime appears almost at the same gas velocity and the transition regime tends to disappear (Majumder, 2008). The efficiency of mass transfer of bubble column is quietly controlled by transport mechanism of species in the liquid viscous boundary layer around the bubbles. The contact time between liquid particles active for the transfer and the bubble, as well as the interfacial area active for the transfer, result in variation of mass transfer efficiency with viscosity of liquid. Mass transfer theories are mainly developed for the process of absorption of a gas into a liquid, even though their application might be extended to the cases that mass transfer occurs between any two immiscible fluid phases. Hence, contact time and diffusion for mass transfer matter for the mass transfer between phases. From the experimental results, it is observed that the mass transfer coefficient depends on the system variables, physical properties of gas and liquid phases significantly. Several investigators correlated their mass transfer coefficient data in terms of dimensionless groups. Therefore, an effort can be made to correlate the entire data of mass transfer coefficient in terms of dimensionless groups by dimensional analysis and the correlation can be expressed as:

$$\left(\frac{k_l}{u_{sl}}\right) = 10.83 \left(\frac{d_c u_{sl} \rho_{sl}}{\mu_{sl}}\right)^{-0.698} \left(\frac{u_g^2}{d_c\, g}\right)^{-0.777} \left(\frac{\mu_{sl}}{\mu_g}\right)^{0.085} \tag{7.23}$$

The ranges of the variables of Equation 8.6 are $0.4 \times 10^{-3} > k/u_{sl} < 1.5 \times 10^{-3}$, $11.45 \times 10^2 > d_c u_{sl} \rho_{sl}/\mu_{sl} < 44.12 \times 10^2$, $0.15 \times 10^{-3} > u_g^2/gd_c < 2.35 \times 10^{-3}$ and $0.79 \times 10^2 > \mu_{sl}/\mu_g < 1.02 \times 10^2$. The correlation coefficient and standard error of Equation 7.23 are 0.978 and 0.083, respectively.

7.5   MASS TRANSFER EFFICIENCY BASED ON MIXING

Mass transfer is an important component of transport phenomena in multiphase reactor operation. For mass transfer phenomena in gas–liquid or gas–liquid–solid systems, the quality of mixing and the liquid side mass transfer coefficient are considered as the most important transport properties. The liquid side mass transfer coefficient incorporates the effects of the liquid flow field surrounding rising gas bubbles. The quality of mixing is very important to determine the efficiency of the bubble columns. The most common approach in treating liquid mass transfer is to combine the mass transfer coefficient and quality of mixing over the entire column height. Earlier works (Majumder, 2008; Manish and Majumder, 2009) analysed the quality of mixedness using information entropy theory in a gas–liquid two-phase flow. In the gas–liquid–solid three-phase downflow, the mass transfer efficiency based on the quality of mixing can also be analysed by information entropy theory (Sivaiah and Majumder, 2013). The gas-interacting downflow slurry bubble column is operating with liquid–solid slurry and gas (atmospheric air). The atmospheric air is flowing concurrently and homogeneously as dispersed phase of bubble in the flow of continuous liquid–solid slurry without reaction. The mass transfer rate can be increased by passing gas through the liquid–solid slurry. If the gas concentration is changed, the rate of increase in mass transfer will be different at different locations of the column. At a complete mixing condition in the column, the liquid phase and the mass rate can be expressed as

$$\int_{C_{l,h_{i-1}}}^{C_{l,h_i}} \frac{dC_l}{(C_l^* - C_l)} = k_l a \int_{t_{i-1}}^{t_i} dt \tag{7.24}$$

On integration of Equation 7.24, one can write

$$\frac{C_{l,h_i} - C_{l,h_{i-1}}}{C_{l,h_i}^* - C_{l,h_{i-1}}} = 1 - \exp(-k_l a t_{i,i-1}) \tag{7.25}$$

where $k_l a$ is the volumetric mass transfer coefficient. The equilibrium concentration $(C_{l,hi}^*)$ in the liquid phase as function of gas–liquid–solid mixing height in the column can be represented as

$$C_{l,h_i}^* \chi = P_{h_i} y_{h_i} \tag{7.26}$$

where χ is the Henry's law constant and P_{hi} is the pressure at different axial locations of the column. This pressure is a function of hydrostatic pressure (P_H), total column pressure (P_t) and the gas–liquid–solid mixing height (h_i) which can be written as

$$\frac{P_{h_i}}{P_t} = [1 + \gamma(1 - h_{r,i})] \tag{7.27}$$

where γ is the ratio of hydrostatic pressure to the total column pressure and $h_{r,i}$ is the dimensionless distance h_i/h_m. The total column pressure and γ can be calculated by Equations 7.28 and 7.29, respectively

$$P_t = P_{atm} + \rho_{sl} g h_i + 4\sigma_{sl} / d_b \tag{7.28}$$

$$\gamma = \frac{1 - \varepsilon_g}{P_r + h_{r,i} + \phi_b} \tag{7.29}$$

where $P_r = P_{atm}/(\rho_{sl} g h_m)$ and $\phi_b = 4\sigma_{sl}/(d_b \rho_{sl} g h_m)$. At constant gas-phase molar concentration, the equilibrium concentration can be measured by Equation 7.30 which can be obtained by substituting Equation 7.27 in Equation 7.26

$$C_{l,h_i}^* \chi = P_t[1 + \gamma(1 - h_{r,i})] y \tag{7.30}$$

The solution concentration and saturation concentration are different in the liquid at different axial locations of the column. Therefore, the local mass transfer efficiency at a particular axial distance from the bottom of the column can be obtained as

$$\eta_{h_{i-1}}^{h_i} = \frac{C_{l,h_i} - C_{l,h_{i-1}}}{C_{l,h_i}^* - C_{l,h_{i-1}}} \tag{7.31}$$

From Equations 7.25 and 7.31, the local mass transfer efficiency at any gas–liquid–solid mixing height h_i can be represented as:

$$\eta_{h_{i-1}}^{h_i} = \frac{C_{l,h_i} - C_{l,h_{i-1}}}{C_{l,h_i}^* - C_{l,h_{i-1}}} = 1 - \exp(-k_l a t_{i,i-1}) \tag{7.32}$$

If the bulk average concentration of the liquid at inlet and outlet of the column are $C_{l,h,in}$ and $C_{l,h,out}$ respectively, then the mass transfer efficiency of the entire reactor volume of gas–liquid–solid mixing height h_m can be calculated as

$$\eta_{bc} = \frac{C_{l,h_m} - C_{l,h_{in}}}{C_{l,h_m}^* - C_{l,h_{in}}} = 1 - \exp(-k_l a t) \tag{7.33}$$

The mass transfer efficiency of the entire column varies with the quality of mixedness at different slurry concentration (Sivaiah and Majumder, 2013). As the quality of mixedness increases, the mass transfer efficiency of the column increases. A functional relationship has been developed for mass transfer efficiency as a function of quality of mixedness ($M(t)$) as (Sivaiah and Majumder, 2013)

$$\eta_{bc} = (m u_{sl}^n) \exp((a u_{sl}^2 + b u_{sl} + c) M(t)) \tag{7.34}$$

where m, n, a, b and c are the parameters varied with slurry concentration as follows

$$m, n, a, b, c = C_1 w^3 + C_2 w^2 + C_3 w + C_4 \tag{7.35}$$

where C_1, C_2, C_3 and C_4 are constants. The ranges of these constants are: $-16.77 \times 10^2 < C_1 < 2.18 \times 10^2$, $-4.20 \times 10^2 < C_2 < 30.74 \times 10^2$, $-15.87 \times 10^2 < C_3 < 2.28 \times 10^2$ and $-0.29 \times 10^2 < C_4 < 2.07 \times 10^2$. The quality of mixedness

$(M(t))$ is a measure to express the degree of asymptotical approach to the equilibrium state and it can be expressed as (Nedeltchev et al., 1999)

$$M(t) = \frac{H(t) - H(t)_{min}}{H(t)_{max} - H(t)_{min}} \tag{7.36}$$

where the values of quality of mixedness $(M(t))$ are varied from 0 to 1. The information entropy $(H(t))$ of the discrete random time variable t $(t_1, t_2 \ldots t_n)$ is defined as

$$H(t) = \sum_{i=1}^{n} P_i(t) I_i(t) \tag{7.37}$$

where $P_i(t)$ and $I_i(t)$ are the probability of the tracer appearing in any semi-cylindrical cell and the information amount, respectively. The values of $P_i(t)$ and $I_i(t)$ can be determined from the experimentally measured values of tracer concentration $(C_i(t))$ as follows

$$P_i(t) = \frac{C_i(t) V_i}{\sum_{i=1}^{n} C_i(t) V_i} \tag{7.38}$$

$$I_i(t) = -\log(P_i(t)) \tag{7.39}$$

where V_i is the volume of the semi-cylindrical cell (Majumder, 2008; Manish and Majumder, 2009).

7.6   PREDICTION OF MASS TRANSFER EFFICIENCY

The factors which conceivably affect the mass transfer efficiency are superficial velocities of gas and liquid, viscosities of slurry and gas, surface tension of the liquid–solid slurry and bubble diameter. The independent effect of all these variables on the mass transfer efficiency is very complicated. Thus, a correlation has been developed by dimensional analysis using Microsoft Excel in terms of geometric, dynamic and physical variables of the system. By multiple regression analysis, the following functional relationship is obtained

$$\eta_{BC} = 25.11 \times 10^{-2} \left(\frac{d_c u_{sl} \rho_{sl}}{\mu_{sl}} \right)^{1.559} \left(\frac{u_{sl} h_m}{E_z} \right)^{0.149} \left(\frac{\mu_{sl}}{D_l \rho_{sl}} \right)^{-1.540} \left(\frac{u_g^2}{g d_b} \right)^{-0.014} \left(\frac{g d_b^3 \rho_{sl}^2}{\mu_{sl}^2} \right)^{-0.378} \tag{7.40}$$

The correlation coefficient (R^2) and standard error of Equation 7.40 are found to be 0.985 and 0.063, respectively. The developed correlation satisfies within the range of variables: $10.93 \times 10^2 < (d_c u_{sl} \rho_{sl}/\mu_{sl}) < 49.03 \times 10^2$, $0.06 \times 10^2 < (u_{sl} h_m/E_z) < 1.19 \times 10^2$, $1.18 \times 10^2 < (\mu_{sl}/D_l \rho_{sl}) < 2.64 \times 10^2$, $0.85 \times 10^{-3} < (u_g^2/gd_b) < 23.64 \times 10^{-3}$ and $0.60 \times 10^{-5} < (gd_b^3 \rho_{sl}^2/\mu_{sl}^2) < 42.88 \times 10^{-5}$.

KEYWORDS

- chemical method
- gas phase
- gas–liquid boundary layer
- mass transfer
- non-pyrophoric Raney nickel
- physical methods
- pyrophoric Raney nickel
- volumetric mass transfer coefficient

REFERENCES

Alper, E.; Ozturk, S. Effect of Fine Solid Particles on Gas-Liquid Mass Transfer Rate in a Slurry Reactor. *Chem. Eng. Commun.* **1986,** *46,* 147–168.

Alper, E.; Wichtendahl, B.; Deckwer, W.-D. Gas Absorption Mechanism in Catalytic Slurry Reactors. *Chem. Eng. Sci.* **1980,** *35,* 217–222.

Álvarez, E.; Gómez-Díaz, D.; Navaza, J. M.; Sanjurjo, B. Continuous Removal of Carbon Dioxide by Absorption Employing a Bubble Column. *Chem. Eng. J.* **2008,** *137,* 251–256.

Beenackers, A. A. C. M.; van Swaaij, W. P. M. Mass-Transfer in Gas–Liquid Slurry Reactors: Review Article. *Chem. Eng. Sci.* **1993,** *48*(18), 3109–3139.

Bin, A. K. Gas Entrainment by Plunging Liquid Jets. *Chem. Eng. Sci.* **1993,** *48,* 3585–3630.

Bin, A. K.; Smith, J. M. Mass Transfer in a Plunging Liquid Jet Absorber. *Chem. Eng. Commun.* **1982,** *15,* 367–383.

Chapman, C. M.; Nienow, A. W.; Cooke, M; Middleton, J. C. Particle-Gas-Liquid Mixing in Stirred Vessels. *Chem. Eng. Res. Des.* **1983,** *61,* 71–95.

Chu, L. B.; Xing, H. X.; Ya, A. F.; Sun, X. L.; Jurick, B. Enhanced Treatment of Practice Textile Water by Microbubble Ozonation. *Process Saf. Environ. Prot.* **2008,** *86,* 389–393.

Danckwerts, P. V. Absorption by Simultaneous Diffusion and Chemical Reaction. *Trans. Faraday Soc.* **1950,** *46,* 300–304.

Deckwer, W.-D. *Bubble Column Reactors;* John Wiley and Sons: Chichester, England, **1992**.

Dietrich, E.; Mathieu, C.; Delmas, H.; Jenck, J. Raney—Nickel Catalyzed Hydrogenations: Gas-Liquid Mass Transfer in Gas-Induced Stirred Slurry Reactors. *Chem. Eng. Sci.* **1992**, *47*(13/14), 3597–3604.

Holstvoogd, R. D.; van Swaaij, W. P. M.; van Dierendonck, L. L. The Adsorption of Gases in Aqueous Activated Carbon Slurries Enhanced by Adsorbing or Catalyst Particles. *Chem. Eng. Sci.* **1988**, *43*(8), 2181–2187.

Jamialahmadi, M.; Muller-Steinhagen, H. Effect of Solid Particles on Gas Hold-up in Bubble Columns. *Can. J. Chem. Eng.* **1991**, *69*, 390–393.

Jin, H.; Yang, S.; He, G. E.; Liu, D.; Tong, Z.; Zhu, J. Gas-Liquid Mass Transfer Characteristics in a Gas–Liquid–Solid Bubble Column under Elevated Pressure and Temperature. *Chin. J. Chem. Eng.* **2014**, *22*(9), 955–961.

Juvekar, V. A.; Sharma, M. M. Chemical Methods of Determination of Liquid Side Mass Transfer Coefficient and Effective Interfacial Area in Gas-Liquid Contactors. *Chem. Eng. Sci.* **1973**, *28*, 976–978.

Kars, R. L.; Best, R. J.; Drinkenburg, A. A. H. The Sorption of Propane in Slurries of Active Carbon in Water. *Chem. Eng. J.* **1979**, *17*, 201–210.

Kikuchi, K.; Loka, A.; Tanaka, Y.; Saihara, Y.; Ogami, Z. Concentration Determination of Oxygen Nanobubbles in Electrolysed Water. *J. Colloid Interface Sci.* **2009**, *329*, 306–309.

Lee, Y. Y.; Tsao, G. T. Oxygen Absorption in Glucose Solution. *Chem. Eng. Sci.* **1972**, *27*, 1601–1608.

Lin, C. S.; Denton, E. B.; Gaskil, N. S.; Putnam, C. L. Diffusion Controlled Electrode Reactions. *Ind. Eng. Chem.* **1951**, *43*, 2136–2143.

Lindner, D.; Werner, M.; Schumpe, A. Hydrogen Transfer in Slurries of Carbon Supported Catalysts (HPO) Process. *AIChE J.* **1988**, *34*(10), 1691–1697.

Majumder, S. K. Efficiency of Non-Reactive Isothermal Bubble Column Based on Mass Transfer. *J. Chem. Eng.* **2008**, *3*(4), 440–451.

Mandal, A. Characterization of Gas-Liquid Parameters in a Down-Flow Jet Loop Bubble Column. *Braz. J. Chem. Eng.* **2010**, *27*(02), 253–264, (April—June).

Manish, P.; Majumder, S. K. Quality of Mixing in Downflow Bubble Column Based on Information Entropy Theory. *Chem. Eng. Sci.* **2009**, *64*(8), 1798–1805.

Marrucci, G. A Theory of Coalescence. *Chem. Eng. Sci.* **1969**, *24*, 975–985.

Meng, F.; Li, X.; Li, M.; Cui, X.; Li, Z., Catalytic Performance of CO Methanation over La-Promoted Ni/Al_2O_3 Catalyst in a Slurry-Bed Reactor. *Chem. Eng. J.* **2017**, *313*, 1548–1555.

Midoux, N.; Charpentier, J. C. Les reacteurs gaz-fiquide a cuve agitee mecaniquement; aires interfaciales. *Entropie* **1981**, *101*, 3–31.

Nedeltchev, S.; Ookawara, S.; Ogawa, K. A Fundamental Approach to Bubble Column Scale-up Based on Quality of Mixing. *J. Chem. Eng. Jpn.* **1999**, *32*, 431–439.

Oguz, H.; Brehm, A.; Deckwer, W. D. *Mass Transfer with Chemical Reactor in Multiphase Systems, NATO ASI;* Alper, E., Ed.; Springer: Netherlands, **1983**; Vol. 2, pp 199–224.

Patil, V. K.; Joshi, J. B.; Sharma, M. M. Sectionalized Bubble Column: Gas Holdup and Wall Side Solid-Mass Transfer Coefficient. *Can. J. Chem. Eng.* **1984**, *62*, 228–232.

Prince, M. J.; Blanch, H. W. Transition Electrolyte Concentrations for Bubble Coalescence. *AIChE J.* **1990,** *36*(9), 1425–1429.

Quicker, G.; Alper, E.; Deckwer, W.-D. Effect of Fine Activated Carbon Particles on the Rate of CO_2 Absorption. *AIChE J.* **1987,** *33,* 871–875.

Ramesh, K. V.; Raju, G. M. J.; Murty, M. S. N.; Sarma, B. C. Wall-to-Bulk Mass Transfer in a Two-Phase Upflow Bubble Column with a Composite Promoter. *Ind. Chem. Eng.* **2010,** *51,* 215–227.

Ravoo, E.; Rotte, J. W.; Sevenstern, F. W. Theoretical and Electrochemical Investigation of Free Convection Mass Transfer at Vertical Cylinders. *Chem. Eng. Sci.* **1970,** *25,* 1637–1652.

Ruthiya, K. C.; van der Schaaf, J.; Kuster, B. F. M.; Schouten, J. C. Mechanisms of Physical and Reaction Enhancement of Mass Transfer in a Gas Inducing Stirred Slurry Reactor. *Chem. Eng. J.* **2003,** *96,* 55–69.

Ruthiya, K. C.; van der Schaaf, J.; Kuster, B. F. M.; Schouten, J. C., Similar Effect of Carbon and Silica Catalyst Support on the Hydrogenation Reaction Rate in Organic Slurry Reactors. *Chem. Eng. Sci.* **2005,** *60*(22), 6492–6503.

Schumpe, A.; Saxena, A. K.; Fang, L. K., Gas/Liquid Mass Transfer in a Slurry Bubble Column. *Chem. Eng. Sci.* **1987,** *42*(7), 1787–1796

Sharma, M. M.; Mashelkar, R. A. Absorption with Reaction in Bubble Columns. *Inst. Chem. Eng. Symp. Ser.* **1968,** *28,* 10–21.

Sivaiah, M.; Majumder, S. K. Mass Transfer and Mixing in an Ejector-Induced Downflow Slurry Bubble Column. *Ind. Eng. Chem. Res.* **2013,** *52*(35), 12661–12671.

Sridhar, T.; Potter, O. E. Interfacial Area Measurements in Gas-Liquid Agitated Vessels. *Chem. Eng. Sci.* **1978,** *33,* 1347–1353.

Tinge, J. T.; Drinkenburg, A. A. H. The Enhancement of the Physical Absorption of Gases in Aqueous Activated Carbon Slurries. *Chem. Eng. Sci.* **1995,** *50*(6), 937–942.Verma, A. K.; Rai, S. Studies on Surface to Bulk Ionic Mass Transfer in Bubble Column. *Chem. Eng. J.* **2003,** *94,* 67–72.

Vermuelen, T.; Wllhams, G. M.; Langlo, G. E. *Chem. Eng. Prog.* **1955,** *51,* 85F.

Vinke, H.; Hamersma, P. J.; Fortuin, J. M. H. Particle-to-Bubble Adhesion in Gas–Liquid–Solid Slurries. *AIChE J.* **1991,** *37*(12), 1801–1809.

Vinke, H.; Hamersma, P. J.; Fortuin, J. M. H. Enhancement of the Gas-Absorption Rate in Agitated Slurry Reactors by Gas-Adsorbing Particles Adhering to Gas-Bubbles. *Chem. Eng. Sci.* **1993,** *48*(12), 2197–2210.

Wang, T.; Wang, J. Numerical Simulations of Gas–Liquid Mass Transfer in Bubble Columns with a CFD–PBM Coupled Model. *Chem. Eng. Sci.* **2007,** *62*(24), 7107–7118.

Whitman, W. G. Preliminary Experimental Confirmation of the Two-Film Theory. *Chem. Metall. Eng.* **1923,** *29,* 146–148.

Wimmers, O. J.; Fortuin, J. M. H. The Use of Adhesion of Catalyst Particles to Gas Bubble to Achieve Enhancement of Gas Absorption in Slurry Reactor. *Chem. Eng. Sci.* **1988,** *43,* 313–319.

Yang, G. Q.; Bing, D. U.; Fan, L. S. Bubble Formation and Dynamics in Gas-Liquid-Solid Fluidization: A Review. *Chem. Eng. Sci.* **2007,** *62*(1–2), 2–27.

Yasunishi, Y.; Fukuma, M.; Muroyama, K. Wall-to-Liquid Mass Transfer in Packed and Fluidized Beds with Gas-Liquid Concurrent Upflow. *J. Chem. Eng. Jpn.* **1988,** *21,* 522–528.

Zaki, M. M.; Nirdosh, I.; Sedahmed Liquid-Solid Mass Transfer at Vertical Screens in Bubble Columns. *J. Can. Chem. Eng.* **1997,** *75,* 333–338.

van der Zon, M.; Hamersma, P. J.; Poels, E. K.; Bliek, A. Gas–Solid Adhesion and Solid–Solid Agglomeration of Carbon-Supported Catalysts in 3-Phase Slurry Reactors. *Catal. Today* **1999,** *48*(1–4), 131–138.

SUGGESTIONS FOR FURTHER STUDY

Downflow gas-interacting slurry system is one of the interesting multiphase flow systems to intensify the transport process by increasing interfacial area and the residence time of dispersed phase. This book engrossed to some hydrodynamics and mass transfer process related to the conventional bubbly flow system. More investigations on the downflow gas-interacting slurry system are essential for further understanding and the industrial use of the system. Some recommendations for further study are specified as follows:

- There is a lacuna of studies for the details of flow pattern in the gas-interacting slurry system. The investigation is needed on the flow pattern in downflow system of the slurry at different operating conditions. The study is required with a wide range of particle size and particle group for optimization process yield. The transition of the flow pattern is a crucial factor to design the bubbly flow reactor at its optimized operating conditions for better yield of the transport processes. More studies on the stability of the downflow slurry system are required. Analysis of the flow patterns by computational fluid dynamics (CFDs) tools is also suggested for further study.

- The downflow gas-interacting slurry reactor is governed by the insertion or distribution of the gas by a special mechanism. The mechanism of gas distribution depends on the degree of entrainment of gas. The optimized process for the gas distribution in the downflow system is required to study for more details. The proper design of the device of gas entrainment by gas induction from the source which is a primary requirement in this downflow system is essential. The mechanisms of the entrainment of gas are yet not fully understood. The physics of the entrainment by different techniques is to be studied in details with different type and sizes particles. Numerical analysis of the entrainment process of the gas is also required in the slurry medium. The effect of distributor design on the flow pattern and the entrainment efficiency need to be investigated along with modeling.

- The occupation of gas volume in the slurry reactor is an important design parameter which is to be further studied with different operating variables and different particles. The studies on the radial distribution of the gas hold-up with different sizes of particles in downflow system are required. Further study is required for the distribution of the gas hold-up since and its analysis by mechanistic model.

- The frictional pressure drop in three-phase flow is analysed by empirical model. Comprehensive mechanistic model is required to develop to analyse the frictional pressure drop in downflow slurry bubble column. The estimation of drag experimentally is still to be performed by sophisticated equipments. It is required to undertake measurements throughout the column, exactly from the gas distribution point to the bottom zone and over a wide range of column diameters in the slurry system. A spectrum of gas–liquid–solid systems should be covered by a wide range of slip/relative velocity. Further, Reynolds stress, energy estimation of spectra and various space and time correlations are to be developed which will provide a very valuable base for understanding the relations between the flow pattern and the design parameters and validate with the CFDs tool.

- The degree of gas–liquid–solid mixing in the gas-interacting downflow slurry system is to be analysed by the computational tool and validate with experiment. There is a lack of study on the radial dispersion coefficient in the gas-interacting downflow slurry system. The contribution of the jet energy in the phase mixing is to be well-established since the intensity of the mixing in the downflow system by jet ejector is significantly higher than that of the conventional upward slurry bubble column reactor.

- Non-adjustable parameters determining phase dispersion coefficient and the transport coefficient during the downflow of bubble against its buoyancy are functions of gas entrainment, bubble size, jet velocity as well as the function of the physical properties especially viscosity and surface tension of the liquid. Theoretical correlations are solely based on the experimental results and may not be suitable for specific applications beyond a range within which the correlation is made. The correlation sometimes cannot be recommended unless it is based on the measurements and material systems investigated by the same material system using apparatus of the same operating conditions or not. Therefore, mechanistic models are required to formulate and interpret the accurate physics of the transport phenomena of the downflow slurry system. The models which are to be made should be checked by a sensitivity analysis which involves simulating the sensitivity of the target reactor features, such as conversion and space-time, yield to variations to transport processes. Special attention to be given to formulate in case of downflow system because of high interactions of the phase in the system.

- The study on the heat transfer characteristics in the gas-interacting slurry bubble column is yet not understood and its study is very scanty. A detailed study is required with various liquids and with a wide range of temperature and pressure.
- The research with suspended solid or slurry in the downflow system is available but only about some hydrodynamics and mass transfer process. More studies are required with different types of non-Newtonian liquid system.
- Development in the formulation of basic equations for the three-phase turbulent flows in downflow system is still not acceptable. Research is needed for the modeling of interface forces, higher-order correlations among phase hold-ups, phase velocities and turbulent dispersion in the phases in case of downflow of gas-interacting slurry bubble column equipped by ejector for gas distribution.
- The simulation of the downflow slurry bubble column by CFDs tool is yet not reported. The correspondence between the CFD simulation and the real gas–liquid–solid system is not available in the literature. The simulation is recommended to perform in a very wide range of various operating conditions.
- The parametric study and their effects on the hydrodynamics and transport processes are recommended to thoroughly investigate by transient simulation. It is required to analyse the correlation between the conditions of numerical simulation and the experimental measurements.
- In the case of downflow gas-interacting slurry system, the degree of shear rate and turbulence in the vicinity of gas entrainment is much different than a conventional upflow system. Under these conditions, the mechanism of momentum transfer is not well understood.

One of the problems is the boundary condition at the interface. The quantitative effect depends upon boundary conditions. Moreover, in the downflow gas-interacting slurry column, there is possibility of a wide bubble size distribution because of distribution of bubble based on its balance of buoyancy against liquid momentum. The bubble also experiences radial forces. For ellipsoidal or spherical cap bubbles, the drags experienced in the vertical and horizontal directions are markedly different. This difference has not been quantified either experimentally or theoretically. All of the above issues need further investigation. A systematic analysis of this problem is needed particularly when the flow is turbulent and the bubble is non-spherical.

- There is a continuous coalescence, breakup and dispersion in the vicinity of jet plunging region. In addition, bubble–bubble inter-action may change the hydrodynamic characteristics. The mass transfer may result in bubble shrinkage or inflation due to absorp-tion or stripping, respectively. Further, in the case of tall bubble columns, the hydrodynamics and the transport processes may change with height due to change in the hydrostatic head. All these issues need to be studied by CFD simulation.

- The analysis of bubble size distribution, effective interfacial area, gas–liquid mass transfer, the transport phenomena for solid suspen-sion, solid–liquid mass transfer and so forth in the case of down-flow gas-interacting system aided by plunging liquid jet by CFDs are to be focussed further.

- Last but not least, the studies are required considering the problem in real industrial process in the downflow slurry system even with the particle density lower than liquid.

INDEX

Milton Keynes UK
Ingram Content Group UK Ltd.
UKHW022047141024
449569UK00022B/831